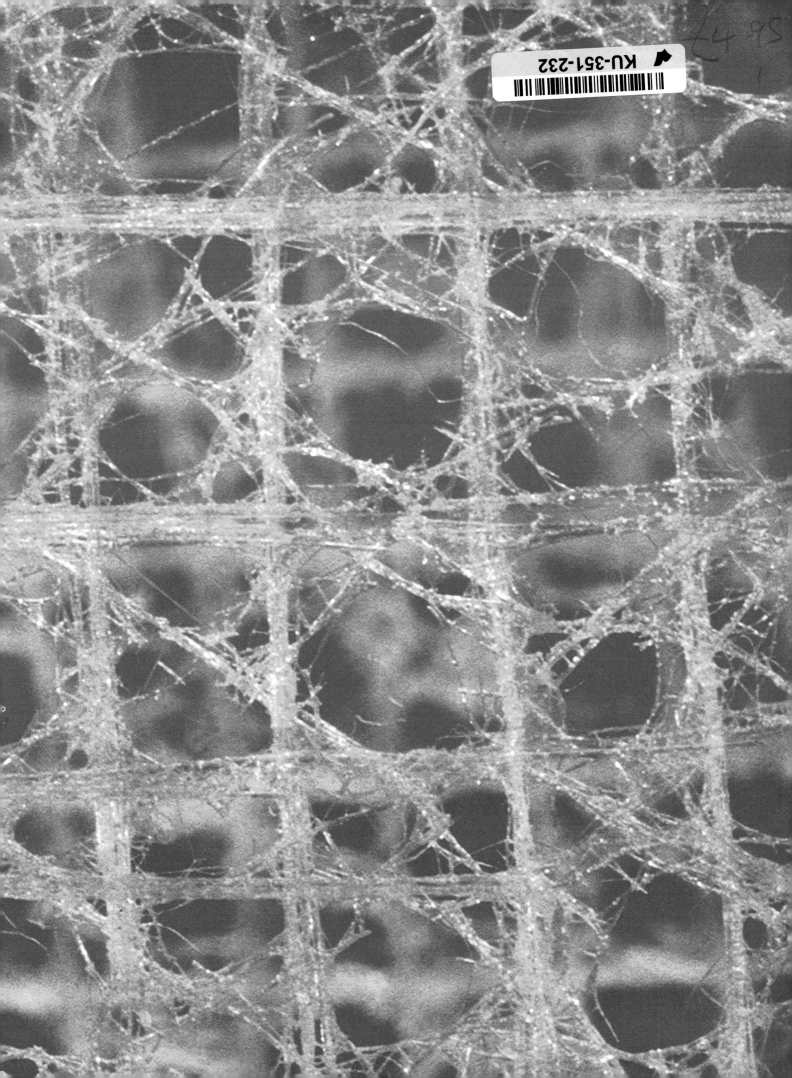

THE GRAND DESIGN

Form and colour in animals

Sally Foy and Oxford Scientific Films
With a foreword by Gerald Durrell

INTRODUCTION BY HENRY BENNET-CLARK
CAPTIONS BY JILL BAILEY

THE GRAND DESIGN

Form and colour in animals

J. M. Dent & Sons Ltd
London Melbourne Toronto

First published 1982
© BLA Publishing Limited and Sally Foy 1982
© Colour transparencies Oxford Scientific Films 1982

Colour origination by Chris Willcock Reproductions
    and Heraclio Fournier SA
Photoset by Southern Positives and Negatives (SPAN),
    Lingfield, Surrey
Printed and bound in Spain by Heraclio Fournier SA

This book was designed and produced
by BLA Publishing Limited, Lingfield, Surrey
for J. M. Dent & Sons Ltd
Aldine House, Welbeck Street, London W1M 8LX.

British Library Cataloguing in Publication Data

Foy, Sally and Oxford Scientific Films
    The Grand Design: form and colour in animals
    1. Camouflage (Biology)
    I. Title
    591. 57′2   QH546

ISBN 0-460-04571-7

END PAPER: Detail of the lower stem lattice of Venus's flower basket.

HALF-TITLE: A queen conch shell stranded in the sunset on a Caribbean
           beach.

TITLE-PAGE: The elegance of animals designed for speed. Probably the
           oldest defence in the world, flight from predators is essential
           for these springboks, which graze in vast herds on the South
           African plains.

Contents

Foreword

IN MY LIFETIME the technology, which assists the professional zoologist to understand how the world works, aids the artist to see new shapes and colours, and the man in the street to appreciate the world around him, has taken enormous strides. I can remember being overwhelmed when, in 1932, someone gave me a pair of thick glass goggles with rubber frames that enabled me magically to 'see' clearly underwater – for as long as my breath held out – for snorkels and aqualungs had not been invented. My first microscope made me realize that, fascinating insect though a cockroach was running about the house, its component parts under the microscope showed you what a miracle of beauty and intricate construction it was – it was the difference between admiring Chartres Cathedral as a piece of architecture from outside, and then going inside with a ladder that enabled you to see every multicoloured fragment of delicate glass in each window. I can remember as a young naturalist my frustration at not being able to see *how* a bird flew, a flea jumped, a flower grew. Now, with new and remarkable techniques we can watch a hummingbird flying backwards in slow motion, can see exactly how a flea performs its miraculous jumps and can watch, like some fantastic ballet, the movements of plants, the unfolding of flower petals.

Obtaining a closer look at animals can only help to understand their behaviour, and to appreciate the ways in which they live. The remarkable photographs and close-up shots in this book should make everyone more aware of the variety of animals and their different features, even the less glamorous ones. The extraordinary work of the photographers of Oxford Scientific Films and the techniques they have developed to reveal the secrets of animal design and functions deserve the highest praise.

This is a new kind of superbly illustrated natural history book, skilfully written to introduce further scientific information about animals and their evolutionary background. I strongly recommend it as a valuable contribution to our understanding of how animals function. I hope its attractiveness will inspire you to an even greater determination to help in the conservation of the whole range of species in the animal kingdom.

GERALD DURRELL
Jersey Wildlife Preservation Trust
Trinity, Jersey, Channel Isles

7

Introduction

HENRY BENNET-CLARK, *Department of Zoology, University of Oxford*

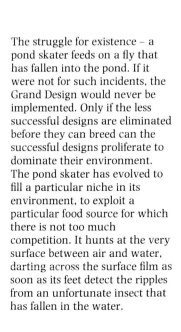

(*Opposite*) Design for living: this small elephant hawk moth is designed principally for flight. Its main purpose is to find a mate and so ensure that another generation of hawk moth is produced. But flight requires a lot of energy, so the hawk moth needs an energy-rich food source – the nectar hidden deep in these flowers. To reach it, the moth has a long flexible proboscis, and its wings are back-swept so that it can hover inside the mouth of the funnel-shaped flowers. As it feeds it is dusted with pollen from the flower which will rub off on the stigma of the next flower it visits. This will fertilize the flower so the plant's survival is also ensured. The similarities in colour between the moth and the flower may not be accidental – from a distance the moth would be difficult for a predator to see. So the shape and colour of both flower and insect are part of the Grand Design.

The struggle for existence – a pond skater feeds on a fly that has fallen into the pond. If it were not for such incidents, the Grand Design would never be implemented. Only if the less successful designs are eliminated before they can breed can the successful designs proliferate to dominate their environment. The pond skater has evolved to fill a particular niche in its environment, to exploit a particular food source for which there is not too much competition. It hunts at the very surface between air and water, darting across the surface film as soon as its feet detect the ripples from an unfortunate insect that has fallen in the water.

The mechanism of evolution

Life is a process by which simple chemical substances are built into complex structures, with the ability to transform one type of energy into another and to reproduce. This process is highly competitive because, while some life forms can exploit non-living energy sources, such as the sun, others rely on the efforts of other organisms.

There are two great divisions of organisms, plants which mostly rely on solar energy to build sugars to power their other life-building processes, and animals which exploit the chemical activities of others both to make the sugars that provide energy for movement and the simple substances they make into their proteins and genes.

This book is concerned with animals. To maintain life, animals must eat. To retain life, they must not be eaten. To eat, they must seek, catch, swallow then digest. To avoid being eaten, they must be aware of danger, be fleet and agile, fierce and dangerous, or simply stronger and more patient than their enemies.

For life is a struggle, in which the fittest survive. The prize, in the struggle, is the ability to breed, passing those attributes that made survival possible to their offspring.

Living organisms are under the continual pressure of competition from other living organisms for food and living space. They can react to this competition in a variety of ways; immediately, by eating or running; gradually, by learning, as a result of exercise, or by growing; more subtly, because they are not identical, each individual's reactions, capabilities and structure differing slightly from those of even his closest relatives. These differences, though, can be the split-second gap between life and death or the slight difference between eating or being eaten.

Thus, gradually, evolution works. And in working, evolutionary pressures act to refine and improve, continually examining the products by the vicious but effective test of survival. In the race, it is not necessary to run twice as fast as the rest, merely to run faster. But in the process of selection, the pressure also acts on the competitors so that

A pair of ithomiid butterflies mating in the Colombian forest, the markings on their wings resembling the interplay of dappled sunlight and leafy shadows among which they live.

what was good enough in one stage of evolution is inadequate or irrelevant later.

The effects of these pressures can be seen in many places, and the fossil record shows that pressures have been acting throughout the history of life on earth. Wherever they live, the organisms that are now alive are experimental survivors that have been tested by present-day animals, plants and the environment. Sometimes the test is brutal and the result dramatic: when man and dog arrived in Australia, many different marsupial species became extinct; when North America re-joined South America after a gap – literally – of forty million years, the indigenous ground sloths, marsupials and ungulates were largely eliminated by competitors arriving from the north; further back, there were massive extinctions of dinosaurs and other reptiles, of brachiopods, trilobites and other groups of animals that failed the test.

But the results of competition in evolution are not always as dramatic. Extinction is normally at the level of the individual, not of the species. Over the time span of life on earth the effects are dramatic. Charles Darwin suggested that such a process could, by trial and error, on several occasions, produce flying organisms, for example, or eyes whose performance approaches the theoretical physical limits. This is hard to believe and leads to questions and doubts, for though the theory of natural selection is able to explain them, some of the stages in the evolutionary process are hard to envisage and evidence for their existence is scanty.

It is striking that the eyes of both the octopus, an invertebrate, and the vertebrates not only look alike but have iris diaphragms, focusing mechanisms, lenses that give a sharp image and external muscles controlling the direction of vision. It is easy to see that these properties are all responses to similar types of selection; that there is for a given role, a particularly good design. Examples of this principle at work in engineering are the wheelbarrow or the bucket of which neither seems capable of great improvement, nor to evolve very fast, though both have been invented using different materials in different parts of the world.

The strange eye of a Tokay gecko can see only in black and white. The gecko hunts mainly by night, so has no great need for colour vision. Photographed in bright light the gecko's vertical pupil has closed down to a narrow slit to prevent the high light intensity damaging its sensitive retina, but a series of small holes remains open. It is thought that light passes through these holes to be focused on the same area on the retina, allowing the gecko to see well even when the light is dim.

An aptly named fairy tern from Cousin Island in the Indian Ocean, perfectly designed for flight. Terns spend most of their lives flying considerable distances over the ocean in search of fish.

Perfect design?

During evolution the animal must survive. Its structures must evolve in a way that allows the animal to compete successfully with others. How then did such a 'perfect' structure as a wing or as elaborate and apparently ridiculous a structure as a peacock's tail evolve?

The bird's wing is clearly advantageous now but it is hard to see how a half-way wing, which was neither a useful hand nor able to support the bird in flight, was an advantage during its evolution. There are not many fossils of early birds so their history is incomplete. It is wrong, however, to view these processes as if they were occurring now – they must be seen in relation to their competitors while they were evolving.

Modern evolutionary thought is able to explain such improbable events using the theory of 'punctuated equilibria'. This suggests that organisms do not evolve very fast so long as the conditions do not change. Since evolution is a process of adaptation to conditions, the organisms, be they plants, grazing animals or predators, will tend to adapt to the climate and each other in a way

that leads to stability. In conditions of this type, organisms that can survive together in a stable way evolve, though their rate of evolution may be slow. When such a system is changed, as can happen with climatic extremes or disease, it sways, but normally returns to its original point of balance. And because the organisms in the system depend heavily on each other, the system may sway disastrously and break down, with extinction of all or many of the species composing it; the survivors will be unspecialized and less dependent on a particular set of conditions or group of other species.

These breakdowns will provide opportunities for population explosions, for adaptive experiment, for radiation into other surroundings. The populations may be small, the rates of evolution, both successful and unsuccessful, may be rapid. Then, gradually, things will settle down again but with different plant and animal species. Such a process leaves few fossil traces of the experiments but extensive fossil records of the periods of comparative stability.

The evolution of the bird wing occurred during a period when there were only a few small mammals, large dinosaurs, other large reptiles including pterosaurs, an extensive insect fauna and tall, tree-

tween the roles of the two sexes. In the vast majority of species, the female lays eggs for which she has provided most of the food reserves. It therefore matters greatly to her that she should give these eggs the best chance by choosing the best mate. Because of the size of the eggs or young that she carries and her need to feed for her young she tends to be more vulnerable than the male. Thus she tends to be an inconspicuous recluse.

How does she choose her mate? Firstly she must ensure he is of the same species, or the union will be sterile! This may be no easy task and leads to the evolution of a courtship which may convey this information. The male has rather different problems; he can mate with as many females as will accept him, with small cost and minor risk. Of course, he must persuade the females to accept him and, if they have a choice, females will favour the male who will be the best father for their young. This is where the sexual selection acts.

A good father may share the care of the young. His constancy may be tested during courtship. Another good potential father may be able to beat other males in combat. He may have the biggest antlers or be the biggest male. Another may be the one who survives predation best but how may this be measured? If he is conspicuous but agile, the argument goes, he must be fit, and therefore the fitter are the more conspicuous – for example, the birds of paradise, the peacock and, near at home, the conspicuous male blackbird.

The evolution of such secondary sexual characteristics can escalate to the Irish elk's antlers (which are discussed in Chapter Five), or the peacock's tail or that of the Siamese fighting fish. There is, of course, a limit when the characteristic becomes so extreme that the male's fitness is impaired or he becomes too easy a prey. Even so, sexual selection seems to work in this way, favouring brilliance of coloration and display in many birds, mammals, fish and insects and, since it affects design at least at the level of surface pattern and structure, is of great interest and importance.

The Grand Design

It is not my aim to describe particular designs in detail – that is the object of the rest of the book. I have tried to show that animals are designed in complex ways and how designs evolved.

As you read on, you will find that the design of animals apparently has been inventive, subtle and refined. It is important to remember that biologists believe that this occurred not by a process of reason and prediction but by trial and error. The sands of time show both the successes and the failures.

This leafy seadragon is not covered in pieces of seaweed – these growths are actually part of its body. It looks more like a drifting seaweed than a fish, which can be a valuable disguise in the shallow weedy Australian waters where it lives.

Basic shapes

Form is all around us. Almost everything that can be discerned by our senses has a form – a shape, an identity by which it can be recognized. Form is what we describe when we distinguish a chicken from an eagle, a dog from a cat, a bull from a cow.

Geometry is the science of form, describing the properties of different shapes at a more complex and abstract level than our senses can define. Geometry is perhaps the most ancient of all sciences. Thousands of years ago prehistoric men made cave paintings showing lines, circles, spirals – geometric abstractions of the natural forms they saw around them. The ancient Greeks played with mathematics and geometry, discovering many of the basic laws that govern the properties of the basic forms. Their laws remain largely uncontested now, some still bearing their names: Pythagoras' theorem, Archimedes' spiral. Geometry has never been included among the so-called 'natural' sciences, but in recent decades its laws have been applied to natural history (the study of living forms) resulting in many fascinating discoveries.

Any visible form must be composed of matter. Matter can be divided into three types, or states: gases, liquids, and solids. Gases, being non-living and usually invisible, are not considered here. Liquids, however, do have a form, and may be composed of material which is, or was, alive.

Most liquids are not sufficiently dense to have a form other than that defined by the shape of their container. But there is one property that helps to give some form to liquids: surface tension. This phenomenon is due to the tendency of molecules near the surface of a piece of water to be attracted to each other, to 'cling together' with the effect of shrinking the surface slightly. If a small body of fluid is not contained – like a drop of water falling through air – the effect of surface tension is to make the drop round: a sphere, the first shape in nature.

Surface tension is one reason for the spherical shape of many microscopic single-celled animals, which consist of a bag of fluid and little more. At a slightly larger scale (though still barely perceptible to the naked eye) is the amoeba, an animal that has no particular shape at all. Here the density of its internal fluid and the outside seawater are about the same, so it floats. It may bunch itself into a sphere, or extend itself into a cylindrical form, or put out pseudopodia – 'false feet'. It does this by flowing: another property of liquids is the capacity to flow. If an amoeba wants to move, it changes its shape. It puts out a 'false foot', then the rest of its body 'flows' after the leg.

The third state of matter, solidity, is the most common in organic forms, but that solidity may vary from the softness of the appropriately-named

(*Above and left*) The myriad shapes of raindrops falling on a leaf are due to the phenomenon of surface tension. This produces a spherical shape which may be distorted by the pull of gravity, by friction with the air, or by collisions with other drops or solid objects.

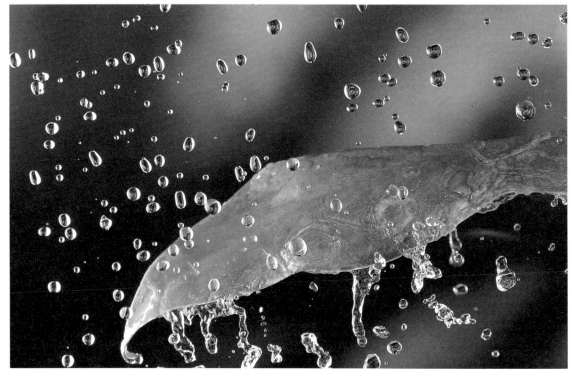

(*Opposite, top left*) The amoeba is a colourless, single-celled organism with a thin and flexible outer gel-membrane enabling it to change its shape easily and to encircle some of its food.

(*Opposite*) Solid objects can be hard and rigid, like this tortoise shell (*right*), or soft and flexible, like the jellyfish (*far right*). Unlike liquids, they can be intricately sculptured.

jellyfish to the hardness of the tortoise's shell. It may be as flexible as hair or as rigid as bone; as deformable as muscle or as strong as an eagle's beak. With solidity comes a multiplicity of forms: shapes, sizes, textures.

However complex the overall form of an animal may be, it can be considered in terms of its component shapes, 'broken down' as a car may be dismantled into its parts. This chapter looks at some of the shapes in nature; how and why they are used by animals; and the skeleton, the basic structure on which everything else depends.

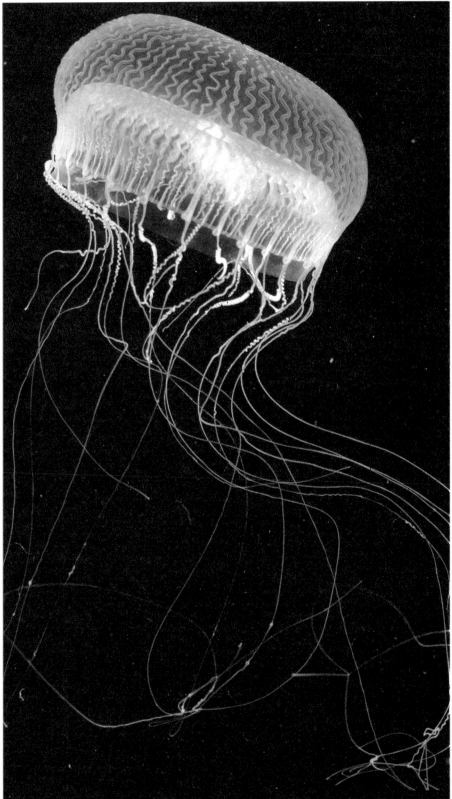

Spheres

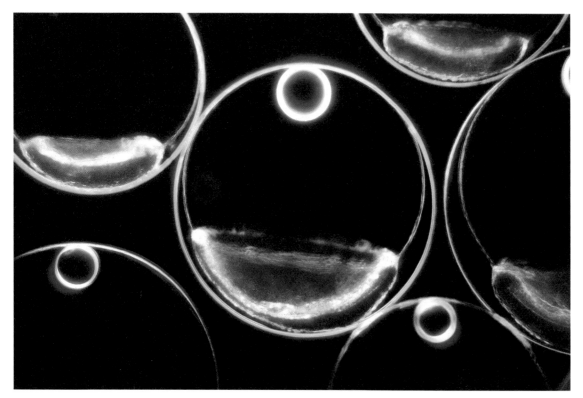

Spheres within spheres. These fish eggs, equally supported by water on all sides, have retained an almost pure spherical form, and so have the tiny oil droplets inside them, used to make the eggs buoyant. Spheres have the smallest possible surface area for their volume, giving a predator less scope for attack.

A sphere has a very special characteristic. It is the shape which presents the least possible surface area for a given volume. This is true whatever the sphere is made of, or whatever size it is. If, for example, a lump of dough is rolled into a ball, it has less surface than if it is shaped into a cube or flattened into a disc; but each way, it still has the same volume.

Surface tension is one reason for a spherical shape to occur, especially in tiny animals composed of more or less fluid matter. At the other end of the scale, in outer space, stars and planets tend to be spherical, since gravity acts rather like surface tension, pulling all parts of the surface to an equal distance from the centre.

Life begins with spheres. The eggs of most creatures, including human beings, are basically spherical. The smallest surface area is the most efficient: not only easier for the female to carry and lay, but allowing least opportunity for predators to break and enter.

Spheres offer little resistance to pressure or friction. If a spherical form always moves in one direction, however, or if the friction of a medium (air, water, or another solid) on a sphere is always in one direction, then the sphere will tend to be distorted in that direction, becoming more oval in shape. This is the genesis of streamlining.

Spheres are also distorted by gravity. If a drop of liquid, held together by surface tension, is placed on a surface and therefore subjected to the force of gravity, it tends to become a more flattened shape, called 'oblate'. The shell of a sea urchin, stripped of its spines, is oblate. This shape distributes stress evenly over the surface and therefore reduces the likelihood of cracking or breaking. The guiding principle of economy is always apparent: a shape is most efficient when it reduces its work to a minimum.

This sea urchin is not quite spherical. Its more flattened shape distributes the stress of gravity over a larger portion of its surface, reducing the strain on the urchin's body, but still leaving it with quite a small surface area to defend against predators.

Who wants to play ball? The small surface area of the sphere is exploited further by the hedgehog. When threatened, it curls up into a ball, its soft underparts and legs tucked safely away while the potential predator faces a sphere of sharp spines.

Most purely spherical forms occur in water where the forces are equal all around the creature. On land, gravity acts with greater effect as the creature gets larger. At an early stage of growth spherical forms start to require the support of beams, struts or trusses of bone and muscle because animals must become more complex in design as they increase in size.

Throughout the animal world, symmetry seems to be selected as a response to various environmental pressures. In most animals the eye has to point in many directions. The eyeball, therefore, is spherical and, bathed in fluid, it moves in a hemispherical socket like a ball in a cup. But the lens or pupil tends to radial symmetry because of the visual axis.

Many larger creatures recognize the value of having the least possible surface area. Rolling into a ball is a simple but effective form of defence, used by creatures as diverse as the woodlouse, the hedgehog, and the armadillo. The economy of shape is made even more effective by adding some form of flexible armour-plating on the surface of the sphere. All the vulnerable and vital organs and limbs are tucked away inside the protective casing, presenting a predator with a frustrating ball game instead of a meal.

The pill millipede has the same strategy: its hard outer skeleton is jointed so it can roll into an impregnable ball, enclosing its head and numerous legs in armour plating. Not only is there no easy way in for the predator, but it would also need a much larger mouth to swallow a rolled-up millipede than a long thin stretched-out one.

Cylinders

The cylinder, or tube as we might call it in everyday terms, is another of nature's more common basic shapes. A hollow form with straight sides and circular ends, it may be made of any material from a liquid to the most rigid solid, and it can be put to a correspondingly wide variety of uses.

Liquid cylinders are almost impossible, for reasons which were mathematically calculated by the biologist Félix Plateau. He showed, by extending soap bubbles between wires, that a cylinder of liquid becomes unstable once its length reaches the measure of its circumference. The greater the viscosity of the liquid, the longer the cylinder may be stretched, but soon the straight sides will become wavy in outline, or unduloid, and after that the cylinder will separate into spherical drops. Imagine trying to paint a wire with liquid varnish: it is impossible to keep the wire evenly coated. The same effect is apparent when a spider builds its web. It coats a strand of silk with sticky liquid from a special gland in its body, thus making a cylinder of liquid around the thread, but the liquid quickly separates into tiny droplets. So, too, does the dew that covers a gossamer web in the morning, forming necklaces of sparkling beads in the grass.

Once the material constituting a cylinder becomes tougher, however, that cylinder becomes enormously versatile. The sea anemone has a simple, muscular, cylindrical body that can be lengthened or contracted at will, but it relies on sea water to support its jelly-like flesh from inside and outside. Nevertheless worms, anemones, or even blood vessels cannot have merely 'liquid' tissue walls but are constrained, within limits, by a fibrous lattice of some kind.

Flexible cylinders make body skeletons which have enormous advantages when it comes to moving around: a considerable volume of body can be passed through a small space – hence the earthworm burrowing through the ground, or the snake slithering through tiny chinks in the rock. As a hollow tube, the cylinder can be used to conduct liquids in or out of small spaces. The mosquito sucks up blood through its cylindrical mouthparts. The elephant's trunk acts as a two-way conduit that can suck water in and blow it out with the force of a

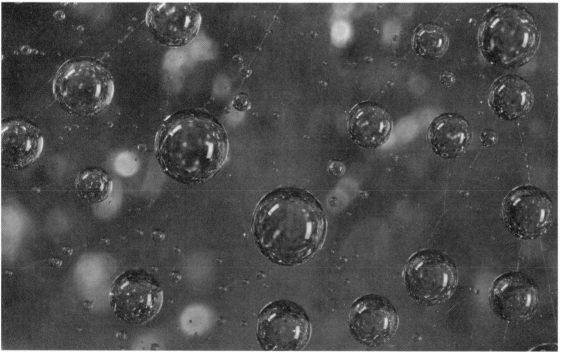

The main spiral of a spider's web is coated in sticky glue to trap flying insects. As the spider extrudes the spiral thread, it is coated with a sticky liquid from the glue glands. Because liquids cannot easily form cylinders, the glue soon contracts to form little beads, visible only with a lens. On a much larger scale, the dew that condenses on a spider's web in the early morning forms strings of sparkling droplets.

The sea squirt's body is made up of two cylinders. Water drawn into the main cylinder through the uppermost opening is filtered to extract food and oxygen, then escapes through the lateral cylinder. The small size of the lateral opening causes a rapid jet of water which carries the sea squirt's waste well beyond the incoming water current.

In spite of its very long abdomen, this damselfly is able to get airborne because its skeleton forms a hollow cylinder around the abdomen, strong but light.

garden hose. Provided the constructive material of a cylinder is flexible enough, the cylinder can be bent round corners, or curled up tightly like a butterfly's proboscis which curls into a spiral when not in use.

Cylinders may also act as pipelines carrying one material through another, like underground pipes. Man's oil or drainage pipelines are usually rigid, but in nature flexibility is more valuable for this purpose. Some bivalve molluscs, such as clams, can live quite deep under the sandy sea bed by virtue of their extensible cylindrical siphons, one for inhaling water and the other for exhaling it after the gills have extracted food and oxygen. In a typical mammal body, the cylindrical arteries and veins which carry the blood have walls, containing the elastic fibre elastin, which expand to accommodate the spurts of blood pumped by the heart and then shrink again, pushing the blood onwards. The heart alone could never propel the blood all the way round the body if the blood vessels had rigid walls – the blood would be stopping and starting all the time, instead of flowing. Strokes occur when the elasticity is lost.

Hollow cylinders exist abundantly in nature. They will support more weight than solid rods of equal length and made of equal amounts of the same material. A hollow beam will resist greater bending or twisting moments. This is immensely useful in the building of skeletons for if the bones of vertebrates were solid and large enough to support their body weight, their skeletons would be far heavier. All the larger supportive bones of the body are hollow. Where lightness and strength are important, as in the wing-bones of birds, the walls of the bones are particularly thin. Cross-sections would show that the walls of a swan's wing-bone, for example, represent only 38 per cent of a solid bone of equal strength. Insect limbs are also a jointed series of hollow cylinders, providing strength with lightness.

Other strong, light cylinders, are useful in making protective homes for vulnerable animals such as the caddis fly larvae. They construct cylindrical cases (disguised with bits of twig or stone as camouflage) which are light enough to carry around like shells, but tough enough to withstand attack.

Radiating shapes

If you let a drop of ink fall on a piece of paper, the splash pattern that results looks rather like a sea urchin. If you drop water into a bowl of liquid and photograph the moment of impact with high-speed equipment, the coronet shape formed at the surface resembles a sea anemone. Yet another of the basic shapes of life – the explosion, or radiating shape – repeats the forms taken by falling drops of water. Radiating shapes occur wherever numerous lines fan outwards from a single central point – whether in a flat plane, as with a starfish, or in three dimensions, as with a sea urchin. The plant kingdom is full of radiating shapes: the majority of flowers have this form, and many plants grow leaves that radiate directly from a stem base; but there are many examples in the animal kingdom as well. Radiating lines, as a construction design, have two useful attributes: they minimize the distance between the centre and the outlying points, and they provide great scope for increasing the surface area of an organism.

The first of these qualities is most convenient in cases where materials must be transported rapidly from the centre to outer points or vice versa. There is a disadvantage, however. If there are a lot of outlying points, the lines tend to become over-crowded around the centre (diagram a). One way to overcome this problem is to develop branching patterns, to reduce the total length of travel and the congestion of lines at the centre (diagram b). If each artery and vein in the body led directly to the heart, for example, the heart would be swamped in a vast tangle of blood vessels. Instead, a few large central vessels divide and redivide into smaller branches. Physically, the resistance to flow or skeletal

strength are reduced when the vessels coalesce or the skeletal rays are fused. Biologically, the smaller branching vessels help animals survive damage and aid their development and growth.

Increased surface area is extremely useful to many creatures. The radiating gills of fish and larval amphibians, like the newt, absorb oxygen from the water over their surface: an increased surface means increased oxygen. The radiating filaments of the fanworm filter the water for food particles: the more water is covered, the more food is found. Many creatures that live sedentary lives, such as sea anemones, have radiating tentacles to collect food from any direction.

Radiating shapes may be used in the interests of defence by presenting the maximum amount of spines around the body to discourage predators. They are used by hedgehogs, sea urchins, and many other creatures. Numerous branches or spikes radiating from the body may also give the impression that a creature is bigger than its actual size: hence radiating shapes may be seen in threatening display, the ruff of the frilled lizard, or with courtship in the splendid tail of the male peacock.

Sea anemones cannot move around much to feed. Instead, they use rings of radiating tentacles to catch passing sea creatures. These tentacles stretch out in all directions to exploit a considerable volume of water. Their shape gives them a very large surface area, which is covered in stinging cells for catching prey.

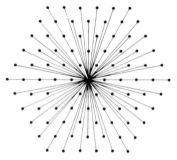

(a) Radiating pattern.

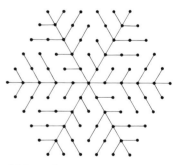

(b) Branching pattern.

(Far left) The 'coronet effect' – as a drop of water falls on to a water surface, a radiating pattern of drops is formed as the surface water is displaced outwards from the point of impact.

Many animals show radial symmetry. For underwater animals the water provides support for quite slender radiating processes, such as sea anemone tentacles *(far right)*, and the tentacles of the jellyfish *Porpita (centre right)*. The sea urchin skeleton *(right)* reveals a body plan based on five groups of radiating structures. In life, these would support spines and tube feet.

The radiating processes ('cerata') of the sea slug *Glaucus*, seen here attacking *Porpita (centre right)*, are filled with stinging cells from its prey. The large surface area of fine filaments is used by the external gills of the newt tadpole *(centre left)* to absorb oxygen from the surrounding water.

A feeding peacock worm *(right)* has a fan of radiating tentacles fringed with fine filaments to sieve food particles from the water currents.

Spirals

Nature's spirals come in all sizes, from the microscopic building-block of life – DNA – to a vast nebula in outer space. On a more mundane scale, spirals may be seen in many plants and animals, from seashells to sheep's horns, from elephant tusks to chameleon tails. Spirals are economical: they can pack a lot of surface area into a small volume of space. They are strong, but flexible. Because they may be unattached at one or both ends, they can be extended easily and endlessly. They are adaptable: they can change their proportions, grow in different directions, or develop into other forms. There are three different kinds of geometric spiral: Archimedes' spiral, the helix and the equiangular spiral.

Archimedes' spiral *(diagram a)* is named after the Greek mathematician who first described it. His spiral starts from a central point and grows outwards in coils of equal width. A tidy sailor coils his rope on the deck in an Archimedes' spiral. The same pattern is used by the orb spider when it makes its web. First the spider makes a radiating pattern of non-sticky silk threads, like the spokes of a bicycle wheel. On these is superimposed an Archimedes' spiral of sticky threads, designed to trap insect prey. The spider itself must be careful to step only on the radiating threads instead of the sticky spiral.

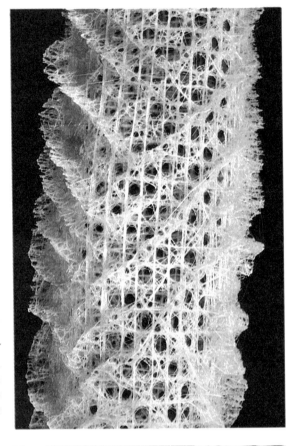

This intricate structure is the skeleton of a glass sponge aptly called Venus's flower-basket. It is made up of rigid 6-rayed silica spicules, secreted by specialized cells, arranged in a highly symmetrical helical pattern.

The sticky trapping thread of the spider's web takes the form of an Archimedes' spiral, with coils of equal width radiating in a single plane from a central point.

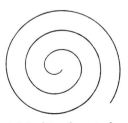

(a) Archimedes spiral.

The proboscis of this colourful 89 butterfly is forming a 'spiral of attitude', a temporary spiral. When the butterfly feeds, the proboscis uncoils and forms a long flexible tube which probes the flower for nectar. The spiral form enables the long proboscis to be 'stored' in a compact way when not in use.

(b) Helix

The temporary spirals of the octopus tentacles may look attractive, but they can rapidly uncoil to seize an unsuspecting passer-by. If the octopus kept its tentacles trailing in the water, moving them towards the prey would create a much bigger warning current of water, and the resistance of the water against the full length of the tentacles would slow down its movement.

The helix *(diagram b)* is a three-dimensional spiral with a constant diameter of curve and a constant distance between coils. The helix is the shape of, for example, a spiral staircase, a corkscrew, or a modern telephone cord. It can be seen in *Euglena spirogyra*, an organism which has both plant and animal attributes. The so-called 'Venus's flower-basket', the beautifully delicate skeleton of a marine sponge, is built on the lines of a helix. The most important helix of all, however, is invisible except under the most powerful of microscopes. It is DNA, or deoxyribonucleic acid, the template that carries the code which informs the sex cells of a creature how to develop into a new generation of the same species. DNA is the fundamental blueprint of life, dictating the design of every animal, and its molecules are arranged in a double helix – one strand coiled round another.

There are two ways to consider the spirals that one sees in animal forms: either as patterns of growth, which are more or less permanent – a particular characteristic of equiangular spirals; or as patterns of attitude, which are temporary. These 'spirals of attitude' occur when a creature arranges itself, or a part of itself, into a spiral shape. An

27

Archimedes' spiral is formed when a millipede curls itself up, or a butterfly coils its long proboscis under its head. A snake at rest may arrange itself in an Archimedes' spiral, placing its head in the middle and curling its body round and round. In all of these cases the principle of economy is obvious – putting lots of surface into a little space.

Few natural patterns are as mathematically precise, or as beautiful, as the equiangular spiral. First described in 1638 by René Descartes, it has several other names: the logarithmic spiral, the geometric spiral, the proportional spiral. It is a pity that none of those names conveys any of the visual appeal of this spiral nor the harmony with which it is incorporated into so many living forms.

The equiangular spiral is essentially a pattern of growth in nature, unlike the helix and the spiral of Archimedes which are mainly seen in attitudes – temporary arrangements of living forms. But unlike most patterns of growth, the equiangular spiral never changes its shape, however large it grows. The radius of each curve increases at a *constant* rate in *proportion* to the last one. The equiangular spiral is a shape of visual and geometrical harmony and constancy.

As they grow, most living things change their proportions and hence their overall shape. For example, a female elephant and her young may look similar, but the mother is not just a magnified version of her young – the proportions of her ears, body, legs, even her trunk, are different. A huge equiangular spiral, however, is just the same shape as a small one. The vast Nebula of Andromeda in space is the same shape as a planktonic snail – though one must be seen through a telescope, the other through a microscope. But there are plenty of equiangular spirals in nature in a scale that we can appreciate without visual aids.

The diagram of an equiangular spiral *(diagram c)*

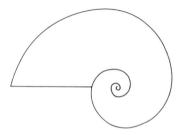

(c) Equiangular spiral.

A living pearly nautilus *(above)* and *(below)* a section through a nautilus shell. The nautilus is almost a living fossil, one of the few survivors of a very ancient and once widespread group of marine animals, which included the ammonites. The animal lives in the last, largest chamber of the shell and looks like an octopus, but has about 200 tentacles. It lives and hunts in the Pacific ocean at depths of over 200 metres, and its heavy shell probably protects it against the great pressure of the ocean at this depth. All the chambers of the shell are connected by living tissue, and each contains both fluid and gas. The nautilus can migrate up and down in the water, changing its buoyancy by pumping liquid into or out of the chambers to change the gas space.

The surface view of a froth of soap bubbles, showing how they tend to form hexagonal shapes. This pattern exposes the maximum surface area to the air, since it requires the minimum amount of liquid soap to be used in the connecting walls (see diagram 4 below).

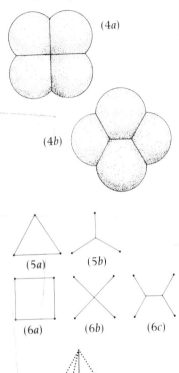

(4a)

(4b)

(5a) (5b)

(6a) (6b) (6c)

(7)

there is a three-way junction o the bubbles meet. Add a four might expect shape 4a to occ

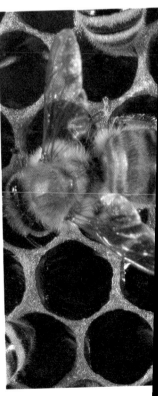

Honeybee workers inspecting cells of the honeycomb. These cells are made by the bees themselves, using wax from special glands on their abdomens. The wax is produced by the young workers, but is hollowed out and moulded by the older workers. These cells are designed to give the maximum volume of storage space for the minimum quantity of wax, and may be used to store honey or pollen, or to house the developing larvae.

The paddle of a fossil ichthyosaur, a marine reptil adapted for swimming. The original bones of the foot h̵ been flattened and reorgan̵ into a single plane to act a̵ paddle. Towards the edge other paddle, in the region whe̵ flexibility is needed, the b̵ are well separated, and n̵ less round in shape. But centre of the paddle the̵ closely packed for stren̵ the more tightly they a̵ packed, the closer their approaches that of a h̵

ong, years
ay
and its
ere
er than other
vs
of present

This picture of the t̵ tortoise's shell show̵ made up of hexago̵ While hexagons ar̵ economical shape f̵ units into flat she̵ surface bec̵ —

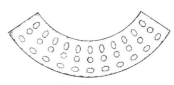

(a) A beam of wood suspended at each end.

(b) A rod of stiff transparent jelly.

like very untidy honeycombs. ̵ct, causing a pressure...̵u can be ̵rms of weight on area – kilograms ̵timetre. A force that pushes is ̵ession; a force that pulls or stretches ̵ion; a force that acts sideways on an ̵ as shear. The action of a force of a ̵ress; the effect it has on the body is ̵e response to stress is measured as ̵d depends on whether the be- ̵. If the material breaks, it does so at ̵nsile) stress – this is its strength. ̵can have very similar short range ̵aviour or stiffness but very dif- ̵Ultimate (Tensile) stress. ̵eton can be seen as a diagram of ̵npression, concepts which may be ̵te simple terms. Imagine a beam of ̵ at each end (diagram a). A heavy ̵d in the middle of the beam. The ̵e beam is pushed, or subjected to ̵he lower part of the beam is ̵jected to tension. As a result of the ̵s strained – it bends, to a greater or ̵ermined by its strength. ̵, that the piece of material being ̵ wooden beam but a rod of stiff

(Far left) Mesosaurus lived about 250 million years ago and was one of the first aquatic dinosaurs to evolve. Evidence from its teeth suggest that it was a fish-eater. It was only about 60 cm long, and its feet were probably webbed to help in swimming. Its foot bones do not have the modifications seen in the paddles of the more advanced marine ichthyosaur (see p.32), which lived about 150 million years ago.

The kangaroo relies entirely on its massive hind legs for movement, and uses its tail as a counterbalance while leaping and also as a 'third leg' when standing upright. Consequently, its vertebral spines are largest over its hindquarters.

(Centre right) The sloth spends most of its life hanging upside-down from branches. Its skeleton therefore has to cope with tension rather than compression. When it leaves its tree, its belly drags on the ground, because its curved spine is designed to support its body weight from below, not from above, and its legs are too weak to support it.

particular. But it is still interesting to note that where the body weight is concentrated the vertebral spines are thickest and tallest. Hence a bison has high spines over its massive forequarters and head, while the kangaroo has vertebral spines over its hindquarters, to support its long powerful legs and tail. The whole bridge diagram is turned upside-down by the sloth, which spends its life hanging from branches: its thin legs are used mainly to resist tension instead of compression, while its spine hangs in a curve, supporting its body weight from underneath.

Greater concentrations of body weight – compression – require larger bones, at the cost of reduced mobility. The elephant has thick leg-bones and is less agile than the bison, which itself is less agile than the deer. A prehistoric dinosaur like

Apatosaurus had its centre of gravity acting through its hindquarters, with massive bones in its hind legs which could take as much as 20 tonnes of its 30-tonne weight at each step. It is unlikely that such a creature could ever have trotted nimbly along; it must have been obliged to shuffle or plod with slow deliberation. In *Apatosaurus*, bison, and most other quadrupeds, the legs are placed four-square under the body to give greatest support. Some reptiles, however – such as crocodiles – have legs that splay out quite widely from their bodies in a posture rather like that of someone doing press-up exercises. To support their bodies they must use tensile muscle strength which costs energy. Not surprisingly, when the crocodile rests it slumps its body on to the ground; it cannot 'stand at ease'.

The legs of the alligator are not far enough under its body for adequate support, so the muscles play a big part in supporting the body. This involves using up a lot of energy maintaining muscle tension, so when the alligator is not moving *(left)*, it does not stand up, but rests its whole body on the ground.

Streamlining

The typical quadruped skeleton is designed to function on land, where gravity is the major external force to be considered. Fish, and other swimming creatures such as whales and dolphins, have a completely different medium to deal with: water. Birds, bats and flying insects must conform to the laws of yet another medium: air. Water and air are both fluids; that is, they flow. The skeletons of both swimmers and flyers have elements of design in common: to accommodate flow they are streamlined. A streamlined body offers the least possible resistance to the flow of water or air.

The average fish skeleton is formed of a few, light bones. The spine must be strong, for almost all the muscles are attached to it via the long vertebral spines. Like the spines of a quadruped backbone, these are longest where stress is greatest, mostly the tensile stress of muscle action. The backbone must be very flexible, for the fish swims by undulating its body and tail from side to side. Limbs disturb water flow, so instead a fish has fine, light, cartilaginous fins, angled to present least resistance to the water flow while helping to stabilize and steer the fish.

Whales and dolphins have taken streamlining

The barracuda (left) shows the streamlined shape typical of most fish. The Hawaiian trunkfish (above), however, has a box-shaped body cunningly patterned to disguise its outline. The trunkfish lives on coral reefs, and does not need to move far through the water, so camouflage is more important than streamlining. The stickleback skeleton (below) shows the flexible backbone and the long vertebral spines which support the swimming muscles.

Since water is much denser than air, streamlining is particularly important for swimming creatures. Most fish are streamlined in shape. There are, of course, exceptions, like the upright seahorse or the box-shaped trunkfish, or fish whose shapes are disguised to camouflage them. But the majority of fish (and most aquatic mammals) are shaped like torpedoes, with rounded heads and smooth, slim, cylindrical bodies tapering away to tails, and few limbs or external appendages. Such a shape will move through water with the least possible drag (diagram a). A body moving through water tends to disturb the flow, causing resistance, or drag, which slows its passage. Drag is created in several ways: friction of the body against the water; variations in water pressure on the body; wave pressure; and turbulence (diagram b) caused by the displaced water swirling in irregular vortices behind the swimming body. The larger a body and the faster it moves through the water, the more turbulence and drag it produces. Therefore any fish that needs to travel fast must incorporate every possible design advantage into its body.

even further than the fishes with a unique mechanism. As they swim, the outer layer of skin 'ripples' in response to slight differences of water pressure over the surface of the body. In this way much surface turbulence is damped and the speed of travel is increased. This phenomenon was first discovered after it had been noticed that dolphins could swim through shoals of bioluminescent plankton, without causing the turbulence that makes those minute organisms sparkle like fireworks in the sea.

Dolphins are streamlined marine mammals whose forelimbs have been reduced to flippers for steering and balancing. The hindlimbs are gone, but the tail ends in two horizontal flukes used for swimming.

The gannet shows a typical streamlined body, with a wedge-shaped head and tapering tail. During flight, the feet are drawn up and back to lie flat along the body. Air-filled sacs connected to the lungs help to keep this large bird airborne.

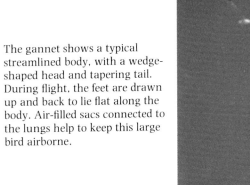

(a) Water flow past a streamlined body.

(b) Water turbulence caused by a swimming fish.

From air to water – the penguin can no longer fly, but is well adapted to aquatic life. Its wings have been modified to form flippers, and its feet are webbed to help in swimming. Its streamlined body is as well adapted to movement in water as it would have been to movement through air.

Birds, are streamlined in a similar way to fishes, since air and water flow have many common properties. The body of a fast-flying bird is shaped very like that of a fish – rounded in front and tapering away at the back. The main difference between living in air and living in water is that while water supports the body weight, air does not, and the heavier a body, the harder it is to become airborne. Several features of the bird skeleton are specially designed for life in the air. The pelvic vertebrae are fused into a solid mass of light bone, the synsacrum, which provides support for the independent movement of wings and legs, and absorbs the compression shock that occurs every time a bird lands on its feet at speed. The tail vertebrae are absent but for a small bone, the pygostyle (find it in the 'parson's nose' next time you eat roast chicken!) which supports the tail. Tiny as it is, that tail controls all vertical, horizontal and fanning movements of the tail feathers, which perform vital steering and braking functions in flight. A special adaptation is evident in the sternum, the bone which links the ribs in front of the chest cavity: it is enlarged and lengthened to

form a deep 'keel'. This provides attachment for the large muscles that lift the wings, and the even larger ones that pull them down again. In ratites – flightless, running birds such as ostriches – the keel is quite small, being no longer needed to support flight muscles.

Not only the shape of the skeleton, but the construction of its bones is adapted for flight. A bird's light, hollow bones are strengthened inside by a mesh of fine struts which criss-cross the hollow cylinders, providing remarkable support. The skeleton of a bird is therefore incredibly light – only 7 per cent of the total weight of an eagle, for example; even its feathers, of proverbial lightness, weigh twice as much. In the interest of achieving lightness, birds have dispensed with heavy jaws and teeth. Their bodies contain buoyant air sacs connected to the lungs which enable their external shape to be controlled independently of the size and shape of the viscera, and of constraints of skeletal shape. To ensure that there is always as much air as possible within the body, a bird's chest muscles are arranged so that it takes effort to empty its lungs, rather than to fill them (as in human beings).

Feathers are a major factor in bird flight design. The plumage serves to protect a bird's body, and keep it warm and dry (see page 106). Hair, however, performs the same functions for mammals, so why don't birds have hair? Although their shape and arrangement are vital to streamline the bird's body outline, feathers can increase body area, especially wing area, at much less weight cost than flesh, bones, or even hair would demand. They have very light tubular spars analogous to those of the wing bones. Feathers are also necessary to give the wings the shape and area needed for flight.

The limits of size

'It was as large as life', we say by way of emphasis. But how large is life? In the animal kingdom, life size may mean a diameter 1/1000th of a millimetre in some single-celled organisms, or a length of 33 metres in the blue whale – the largest animal that has ever lived on earth. Weights range from an infinitesimal fraction of a milligram to more than 150 tonnes.

Despite this almost incomprehensible range of possible sizes, the range of scale within specific animal groups, such as the insects or the reptiles, is much more limited. Within the families – the cat family, for example, or the dog family – the range is still narrower. Finally, within each species there is hardly any variation of adult size. Each species has its own niche in nature, and its size is vital to its long-term tenancy of that niche.

Eggs large and small – from left to right, the enormous fossil egg of an elephant bird, then the eggs of an ostrich, a chicken and a hummingbird. The size of eggs does not depend only on the size of the bird that lays them: a hummingbird looks after and feeds its young in a nest hidden from predators, whereas the chicken lays its eggs on the ground and the chicks must run around and search for their own food as soon as they are hatched.

As we grow up we become accustomed, through education and visual experience, to the 'normal' size of life. Size is the first qualification in most descriptions: 'there's a *big* dog in the garden'; 'there's a *small* frog in the pond'. Having seen plenty of dogs and frogs, we have arrived at a mental image of the size of the average dog or frog. When told that it is large or small, we multiply or divide the average. Only rarely is our assessment of large or small shaken: perhaps when we see a rare animal, like a whale, or when a creature is at the upper or lower size limits of its family, as a moose and a mouse deer are at the opposite limits of deer size. But there are limits. Neither moose nor mouse

deer can afford to grow too large or shrink too small if they are to maintain their hold on life. Not only can they not afford it, they cannot physically manage it while retaining the shape and characteristics of deer.

Size is no accident. It is tailored to many conditions: the effect of external forces – gravity, water pressure, temperature, light, humidity, and so on; the quality, quantity, and availability of food; the number and nearness of predators, kin, and mates. At all times, size is governed by geometric laws that dictate whether an insect can grow as big as an elephant, or why a Shire horse is a different shape to a Shetland pony.

Galileo was the first to point out, around 1600, that strength does not increase in proportion to size: a large structure, built to the same proportions and of the same materials as a smaller one, is the weaker of the two. The larger animal must support

Within any given species, there are definite size limits imposed by the genetic programme contained in its cells, which controls its body design. If the adult giant tortoise in this picture were to grow much larger, its legs would be unable to support the weight of its great shell. If a baby tortoise were too small on hatching, it might be easily taken by a predator, or it might lose body heat too easily.

A single elephant weighs more than a hundred mice, but it has to carry all this weight around as it walks. It is much less likely to be attacked and eaten than a mouse, but it has to travel much further to find enough food to eat. For its size, an elephant cannot run as fast as a mouse, but it does not need to.

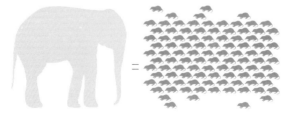

Both the Shire horse *(above)* and the Shetland pony *(below)* are horses, but they contain slightly different genetic information. The Shire horse has been bred for pulling heavy carts and ploughs, and its legs are larger in relation to its body. The Shetland, on the other hand, needs to travel only from one pasture to another, and has smaller legs. The Shire foal's legs are even larger in relation to its body than those of its mother: in the early weeks of its life it will need to be able to keep up with her in order to suckle when it chooses.

The weight of an animal usually increases at the same rate as its volume. At first sight birds would seem to have cheated the size/weight ratio. In fact they show distinct scale effects: above about 1 kg they cannot hover; above approximately 8 kg taking off is very difficult; the absolute upper limit is approximately 20 kg; below about 1 g they can neither keep warm nor beat their wings fast enough.

relatively more weight than the smaller, so its skeleton is built on a different plan. The legs of the Shire horse are proportionally, as well as actually, bigger than those of the Shetland pony. Where the skeletal material cannot adapt, as in the insect cuticle *(see page 70)*, the size of the creature soon reaches its limits. Among water creatures, however, skeletons retain similar proportions as they increase in size, since water supports the animal's body weight.

The ostrich *(left)* and the flamingoes *(right)* show two very different pairs of legs. The ostrich family long ago gave up flight as a means of locomotion and evolved large powerful legs for running and stubby wings used mainly in sexual display and heat control.

Size and metabolism

Toiling for one's daily bread is the lot not only of mankind but of all animals. In order to stay alive, every creature is obliged to eat food and convert it through the body into energy, which is needed to find more food to eat . . . and so on *ad infinitum*, or at least until death. Man, and some other large animals can afford to devote at least part of their time to pursuits other than finding and eating food: a man only needs to eat one-fiftieth of his own weight daily, while a mouse must eat half its weight, and a shrew even more. A shrew is a tiny eating machine facing a constant starvation crisis.

Man and shrew are both mammals. So why is there so much difference in their needs? It all comes down to size. The processes of metabolism (the conversion of food into energy) are similar throughout the animal kingdom. It is all a question of getting 'fuel' or food to the parts of the body that need energy (such as the muscles), and supplying the oxygen needed to 'burn' the fuel and release the energy.

In very small, simple creatures like the flatworms, oxygen diffuses into the creature over its entire surface. There is a limit, however, to how deep and fast oxygen can be absorbed into the body in this way, so such creatures cannot grow very thick or the oxygen will not be able to reach those parts where it is needed. Also, the surface of the creature must be as large as possible relative to its size; so the flatworm stays small, thin and flat, though it may grow longer and wider.

Insects do not have lungs. Their tough external skeletons (*see page 71*) are punctured with rows of tiny holes, or spiracles. These admit oxygen into the body through a system of tubes, or tracheae. Smaller branches of these air-tubes lead directly to the organs where oxygen is needed, such as the flight muscles. These muscles are bathed in liquid food so the oxygen does its work on the spot, releasing the colossal amounts of energy that a flying insect uses – about 1000 calories per gram of muscle per hour, or up to 20 times as much as a

Amid a shower of speedwell petals, a pigmy shrew attacks a worm. An animal as small as a shrew has a large surface area in relation to its size, so it tends to lose heat easily. In order to produce enough heat to remain active, it must consume large amounts of 'fuel' in the form of food, and needs to feed almost around the clock.

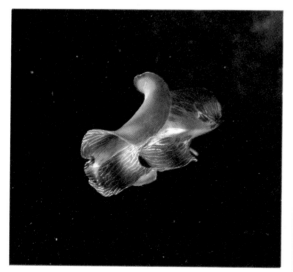

A marine flatworm makes use of its flexible flattened form to swim over a Red Sea reef. As it swims, it absorbs oxygen from the surrounding water over its whole surface.

A paper wasp moulds the cells in which it will rear its young. It obtains its oxygen from the surrounding air through microscopic holes in its outer skeleton. Because insects do not have a very efficient internal transport system for oxygen, their size is limited by the distance over which oxygen can efficiently diffuse to the tissues. This wasp's body is nowhere very thick, so oxygen can reach all the tissues fast enough to allow it to lead a very active life.

This portion of the body of a moon moth caterpillar shows the spiracles very clearly, ringed in red pigment. The spiracles are equally distributed among its body segments, and can be closed if it is in danger of dehydration. Each new skin that it grows will have pores for the spiracles, which connect with a network of tiny tubes that carry oxygen deeper into its body.

running man. Again, there is a limit to how far oxygen can penetrate an insect's body at a sufficient rate, and this is one of the factors that stops insects from becoming as large as elephants.

Moving up the evolutionary ladder we come to the creatures with blood circulatory systems, whose bodies may grow much larger since they have blood to carry oxygen to the body organs. To supply the large quantities of oxygen they need, the oxygen-absorbing surface must be enlarged (the body surface, as we have seen on page 40, does not increase as fast as the volume of the animal). Fishes do this with gills, which are feathered and fringed to provide extra surface area. Reptiles, birds, and mammals have lungs, which are filled with a spongy honeycomb of oxygen-absorbing tissues. If the total surface area of a man's lungs could be spread out flat it would cover about 50m², or carpet the floor of a 5 × 10 metre room.

Fish and reptiles are called 'cold-blooded' since they do not maintain a constant body temperature. Thus they need less food than 'warm-blooded' animals, but they depend on the climate to keep them warm. Fish generally stay about the same temperature as the water they live in; they can go for long periods without food. Reptiles tend to live in warm parts of the world, especially large reptiles

like crocodiles. Lizards lying in the sun can regulate their position and colour so as to maintain a constant input of solar heat and have a body temperature that remains almost the same as ours. This works best for a small animal; a very big one would not cool down very much overnight. If reptiles get too cold they cannot function: snakes in temperate latitudes must hibernate through winter.

Birds and mammals are 'warm-blooded': they keep their bodies at a constant temperature which helps them to metabolize faster – usually about 37°C in mammals, 40°C in birds. As heat will escape from their surfaces, their bodies are usually insulated with a heat-conserving layer of feathers or fur. Because of the disproportionate increase of weight and surface, small bodies with relatively more surface lose heat faster than large bodies with relatively little surface. So the smaller body must metabolize faster to produce more heat than the larger body, just to stay at the same temperature. The smallest mammals, such as shrews, may have metabolic rates that are one hundred times those of the largest, such as whales or elephants. If a cow had the same metabolic rate as a mouse it would probably boil because it could not lose heat fast enough.

Biologist Carl Bergmann, in 1847, formulated what is known as Bergmann's Rule: that among warm-blooded animal species, individuals that live in colder areas will tend to be larger than those that live in warmer areas – by having less surface area for their body mass, they lose less heat. This rule is true of many animals with wide distributions: bears, rabbits, foxes and deer, for example, tend to be larger in colder climates. But the rule has too many exceptions to be considered a useful criterion of size, and it takes no account of different lifestyles and insulation methods by which animals may compensate for size.

The polar bear *(far right)* is one of the largest land mammals, up to 725 kg in weight, while even red foxes living in the far north *(right)* are bigger than their relatives further south. One possible explanation is that larger mammals have a relatively small surface area in relation to their volume, so tend to lose heat less readily than small mammals. But larger animals also have a greater capacity for storing food in the form of fat as a buffer against winter shortage, and the fat also acts as an insulator against heat loss.

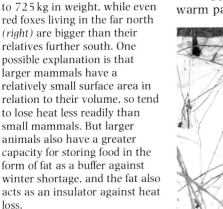

Size and movement

If a rabbit and a horse raced over a short distance, either of them could win. Both are capable of sprinting at about 70 kmph. But the horse would still be running long after the rabbit had collapsed with exhaustion. The horse is about ten times as long and as wide as the rabbit; so, because weight increases as the cube of linear dimensions *(see page 41)*, it is approximately a thousand times heavier. Being ten times bigger it will take ten times as long to make the same movement as the rabbit, but in so doing it will cover ten times as much distance; so they will both arrive at the same place at the same time. The rabbit, however, will have made ten movements to each one by the horse, so the horse should be able to travel ten times farther before being equally tired. But because larger animals metabolize more slowly than small ones *(see page 42)* the horse can keep moving much longer.

In general, large animals can move faster and farther than smaller ones. The relationship between size and movement, however, depends on many factors, and largest does not always mean fastest. A big racehorse will produce more power, and therefore run faster, than a small racehorse; but a carthorse, though it is larger than either of them, cannot compete. An elephant can move surprisingly fast when it is charging, but it is not

built for speed. Its great bulk needs large thick legs to support it, which take a lot of energy to move. It is difficult to see how some of the largest dinosaurs could move at all, but presumably they did. The whale, though larger than most dinosaurs, does not have this problem: water supports all its

The huge tail flukes of a hump-backed whale break water as the animal dives. One of the largest living mammals, the whale uses its tail for swimming. The tail with its horizontal flukes is moved up and down to drive the animal through the water, while its small flippers, the vestiges of its forelimbs, are used for balancing and steering.

A grasshopper in mid-leap is more than its own height above the ground. You can just make out the herringbone pattern on its femur (thigh), which reveals the arrangement of the powerful blocks of muscles inside. These are all attached to a central tendon which is connected to the tibia. Grasshoppers will leap to escape danger, but they also use leaps for getting from one place to another – quicker than walking through the undergrowth.

A millipede advances along a twig. Although renowned for the number of their legs, even the longest millipedes have only about 680 legs, and most species have far fewer. You might expect that an animal with so many legs would move very fast, but the millipede's legs are so short and its fat body so close to the ground that its legs take only short strides at a time. Nevertheless, they can deliver considerable thrust, and millipedes are strong enough to burrow into the ground very efficiently.

weight, leaving its energy free to power its muscles. With a few flicks of its great tail it can surge through the water, while the herring, swimming alongside, must work much harder to gain less overall distance. Speed in water is affected by the surface/weight ratio. The huge whale has relatively little surface area compared with its weight, so water drag is relatively less than for a small fish, and can be easily overcome by its superior muscle power.

On land, some creatures can achieve running and jumping records that belie their small size. A flea may jump 30 cm, two hundred times its length; the grasshopper may leap nearly a metre, twenty times its own length, or the equivalent of a man clearing half the length of a football pitch at one bound.

The flea relies on energy storage in an elastic pad in the side of the body. In locusts there are springs in the hind legs. In kangaroos, the bounce is produced by energy storage in the Achilles tendons. The more energy an animal can produce and *release* into the back legs the higher it will jump. Clearly the height which a flea and a grasshopper can reach also depends on their small size and consequent low weight. Limb design is important and such animals have long hind legs shaped like spring levers.

The leg movement of the millipede occurs in a wave along the body: certain groups of legs are movng forwards as others are thrusting backwards. At any given time there are always some legs in contact with the ground at intervals along its body.

In running, one of the factors contributing to speed is how often the legs of the animal must touch the ground; in other words, the length of its stride. Every time the legs hit earth, its speed is checked for a fraction of a second. If this energy can be stored, as in tendons, the animal can reuse it. The body keeps moving forward through the air by momentum but the legs must use an extra spurt of energy to keep up, as it were. To capitalize on this body momen-

(*Far left*) A cheetah chases a Thomson's gazelle. The two animals show different phases of the running sequence. The gazelle has thrust off from the ground with its hind legs and is at full stretch, covering as much ground as possible before touching down for another thrust. The cheetah has just touched down with its forefeet and is bringing its powerful hind legs forward for another thrust. Note how it holds its tail out behind as a counter-balance.

(*Left*) A bobcat finally catches up with the snowshoe hare it has been chasing. The cat family rely on getting close enough to their prey during a chase to slash with their claws at the prey's hindquarters. The braking effect of this, together with the loss of blood, usually slows the prey sufficiently for the cat to catch up with it and kill it by a bite to the neck.

tum, fast animals tend to be built with their body bulk set as high as possible over long, thin legs, which can be moved with the minimum energy expenditure.

Extra legs do not help an animal to move faster. The millipede is slow for all its legs – in fact, if it hurries it is liable to trip over its own feet! Insects have six legs and tend to have three of them on land at any given moment while moving; they can therefore stop dead without falling over. A four-legged animal in motion may have only one or two legs on the ground at any moment, which makes it much less stable; three legs on earth ensures stability. But two legs can be very efficient. Some lizards bring up their forelegs and sprint on their hind legs only. Ostriches can reach speeds of about 80 kmph, faster than any other two-legged runners.

As we have mentioned, when an animal runs fast there are moments when its four legs are all off the ground at once. During these split seconds there is nothing but air drag to slow down the forward travel, or momentum, of the body. Therefore the longer the stride through thin air, the faster the

animal tends to move. The larger the animal, the more momentum it has once it is moving; this keeps it going forwards faster, but makes it less easy to brake or turn suddenly. (This is the basic premise of bullfighting.) Like oil tankers at sea or aeroplanes landing, large animals need more distance to lose momentum. By the same token they need more space to get going. Big birds, like swans, must 'taxi' along an earth or water runway to gain take-off speed. Even larger birds, like condors or albatrosses, can hardly lift themselves from the ground by wing power alone: they prefer to launch themselves from cliff faces into the wind and to use air pressure or currents to take them aloft. Small birds, on the other hand, can effectively achieve vertical take-off, allowing much greater manoeuvrability.

A small size is of definitely advantageous for some forms of movement other than sheer speed. Digging animals can dig faster and cover themselves more quickly if they are small. Small climbers can move more nimbly than large ones, and use thinner branches with impunity. Animals that glide, like 'flying' squirrels, must be small and light enough to glide, rather than plummet to earth.

The cheetah is claimed to be the fastest mammal on earth, reaching speeds of up to 70 kmph, but it cannot keep up this speed for more than about 400 metres. It relies on a short sharp sprint to catch its prey. Its whole body is adapted for high speed. It has a small light skull, and there is no surplus fat on its streamlined body. Its whole skeleton is remarkably flexible: the backbone curves alternately up and down during running, and the shoulders and hips swivel in relation to it. The long tail is used as a balancing device to stabilize high speed sprints.

An English robin is making almost a vertical take-off as it starts to fly away. It has pushed off from the ground with its legs, and is starting to open its wings. The powerful slow downstroke of the wings will provide most of the lift for take-off, and the upstroke will provide most of the thrust. Small birds like robins can take off in very confined spaces, such as this gap in a bramble patch.

Large birds like swans need a lot more room to get airborne. They use a technique similar to that of aeroplanes, turning into the wind and running over the surface of the water, flapping their wings until they gain enough airspeed for lift-off.

Size and parasitism

There is one career in the animal world for which size is a prime qualification: parasitism. Parasites generally have to be small in order to succeed; and very successful some of them are, too. Parasites are animals that depend for their food (and sometimes also their lodging and care) on other animals, usually without killing them. The latter are known as 'host' animals, although that term implies a social invitation to dine which is not generally extended – the parasite is a gatecrasher.

The smallest living organisms of all, the tiny viruses, are parasites, unable to replicate outside the body of their host. Viruses are so small that they can only be seen through a very powerful microscope. Only a little larger, and often also parasitic, are the bacteria. These may occasionally be beneficial to their host – for example, most plant-eating animals cannot digest cellulose, the tough covering of plant cells, so the job is done for them by colonies of bacteria living within their guts. In such cases the term 'parasitism' is dropped and a more polite word – mutualism, or association for mutual benefit – is used. Most bacteria are no so kind. Viruses and bacteria cause some of the most deadly diseases of man and animals and, in looking after their own interests, as it were, have been responsible for wiping out whole populations and civilizations.

Wholesale destruction is not, however, the usual intention of a parasite, for in destroying its hosts it may destroy its own livelihood. This is where smallness comes in. By being small in relation to its host, a parasite can settle on a creature which apparently has resources to spare. The parasite imposes a tax that the host can afford, but which is still a burden. High levels of parasitization often reduce viability and/or reproductive capacity. The advantage of smallness may also be negated by sheer numbers. If too many parasites attend one host they may weaken it so much that it can no longer support them. Tiny hookworms, parasitizing man and other animals, may reproduce until some five hundred individuals are living in one person, between them taking about 250 ml of his blood daily. A healthy person can make up this loss, but of course the people who are vulnerable to such parasites are not generally in the best of health, and in such numbers hookworms can kill the elderly, infirm, or undernourished.

Flatworms and roundworms are often parasitic and can develop into the largest of internal parasites. Tapeworms that live in human beings may grow as long as 9 m; cattle tapeworms are capable of reaching 15 m, although they are still only 2 mm wide. Parasitic worms often have

A rabbit flea is caught in the act of biting a rabbit's ear. Fleas have sharp biting mouthparts and feed on the blood of their host. Unlike many parasites, they do not need to live on their host most of the time – they just need to visit it for meals. But the rabbit flea is often itself parasitized by another organism, the myxomatosis virus. This has apparently no effect on the flea, but when it bites the rabbit, the myxomatosis enters the rabbit, with often fatal results. In the Middle Ages, fleas that fed on man were the vectors of bubonic plague. Fleas are wingless, leaping from host to host using their powerful back legs.

(*Far right*) The colourful red object looking slightly out of place on this python's body is actually a tick. Ticks also feed on blood, and have specially flattened bodies adapted for crawling under scales, feathers or fur. They can take in so much blood at one meal that they need only three meals in a lifetime, and when fully gorged with blood they drop off their host.

several larval stages, some relatively harmless, others parasitizing different creatures on the way to their final host. Transfer from one host to another often occurs when a scavenger eats an infected dead animal. The parasite must stay small through-out these early stages, for if it is large enough to be seen it may also be avoided. Once it reaches its final resting place it grows rapidly – a tapeworm may be only a few millimetres long when it enters a man, but can grow to a metre within a few weeks, and live for years if not disturbed.

Most internal parasites have reduced some of their body organs and functions to a bare mini-mum. They tend to have no mouths, just hooks or suckers to cling to their host; they absorb food through their body surface, which is extensive in relation to their size: thus they have no need for, and have dispensed with, digestive organs. Since the chances of meeting others of their kind within the depths of their host's intestines are slim, they are often hermaphroditic – they contain both male and female sex organs, and can fertilize themselves. Their eggs – extremely small and therefore difficult to detect and destroy – are passed out in the animal's waste matter, to start other colonies.

All the parasites mentioned so far live inside the bodies of their hosts, but external parasites ('ecto-parasites') are more familiar to us; many of them are quite large enough to be seen. They produce a local anaesthetic and their attentions are often noticed too late by the itching skin they cause. Most of them are bloodsuckers. They include leeches, worm-like creatures that are rife in the humid tropics; lice, mites and ticks; and insects such as flies, mosquitoes and fleas. Ectoparasites are often

The red lumps on the feet and legs of this southern lubber grasshopper are not part of its colouring, but are the nymphal stages of the velvet mite. The nymphs hatch from eggs buried in the ground, then attach themselves to a grasshopper, driving their knife-like mouthparts through its cuticle to suck blood. In a few days they are gorged with blood and drop to the ground. The adult mites are not parasitic, but feed on insects.

This trout looks as if it is feeling the effects of the river lamprey firmly attached to its side. The lamprey uses a sucker-like jawless mouth to cling to the trout, and has a vicious abrasive tongue with which to gorge a hole in its host's body. There it laps up the body fluids oozing from the wound. Lampreys used to be more numerous than they are today: King John died of a surfeit of lampreys, and relatives of the lamprey once wiped out the entire trout population of one of the American Great Lakes.

(*Below left*) A hedge sparrow foster parent is feeding a young cuckoo several times its size. So strong is the bird's instinctive response to the gaping orange mouth that it continues to feed the giant youngster even though it has to stand on the cuckoo's back to reach its bill. The hedge sparrow is quite emaciated from the strain of collecting enough food for its 'adopted' young. (*Below right*) A baby cuckoo is attempting to throw its foster brother out of the nest. Already it has hoisted the young hedge sparrow on to its back, and next it will throw out the remaining blue hedge sparrow egg. By the time the young cuckoo is half grown it will completely fill the little hedge sparrow nest.

Red mites feeding on a harvestman. Not only does the harvestman provide a source of food for the mites, but it may also carry them a considerable distance, helping them to colonize new areas.

mobile and may be non-parasitic for periods of their development. They may inhabit many hosts for short periods before multiplying and moving on (like fleas) or they may live independently of their hosts except when they stop for a meal (like mosquitoes). Among the latter type are the largest parasites of all, the lampreys, eel-like fishes with sucker mouths by which they attach themselves to a larger fish and suck its blood. Only some species of lampreys are parasitic, and among these only the adults are parasitic; but these may reach more than a metre in length and cause severe wounds to the host fishes. (Adults of other lamprey species do not eat at all. They meet to mate, and then die.)

The relatively large size of most mammals precludes them from leading a parasitic lifestyle, although 'vampire' bats have achieved a notorious success in this field. Some birds are semi-parasitic in that they use other birds to incubate their eggs and rear their young. The cuckoo is a well-known example. It lays an egg in the nest of the host bird – an egg which is often cleverly matched in colour and in size to those of the host. The young cuckoo usually hatches first and tips out any other eggs in the nest. Having the home to itself, it then keeps its foster-parents extremely busy providing large amounts of food, while it grows fast until it is two or three times their size.

There is some kind of moral revenge in the fact that parasites are not themselves immune to parasitism. It was Jonathan Swift (1667–1745) who noted:

'So, Nat'ralists observe, a Flea
Hath smaller Fleas that on him prey,
And these have smaller Fleas to bite 'em,
And so proceed *ad infinitum.*'

Larger parasites do, indeed, often carry smaller ones, down to a smallness that can only be appreciated through a scanning electron microscope. This is a world as strange as any in science fiction, where weird creatures inhabit crevices on the bodies of animals that are themselves so small that we can hardly see them. Such chains of dependence are known as hyperparasitism. The vampire bat may be bitten by flies which are bitten by mites which are fed on by bacteria, and so on. The taxman is not spared being taxed.

51

Sexual dimorphism

The male deep sea anglerfish is truly attached to his mate. Once he has found her, he never lets her go; theirs is a marriage bond that lasts for life. He is not put off by the fact that she is large and dangerous – after all, in the eternal midnight of the ocean depths where they live, he cannot see her.

The male and female anglerfish provide one of the most extreme examples of sexual dimorphism, or difference of appearance between the sexes. Living in the sea at depths where light never penetrates, it is not easy for a fish to find a mate of its own species. The female anglerfish is squat and lumpy in shape, equipped with many needle-sharp teeth in her huge mouth, and a small luminous lure like a fishing-rod (hence *her* name) by which she attracts prey fish within reach of her jaws. She may be anything from a few centimetres to a metre in length. The male, by contrast, is extremely insignificant. As a small free-swimming immature fish he somehow finds a female and attaches himself with his mouth to her body. From then on he sticks with her; his mouth becomes fused into

Taken for a ride – a tiny male deep sea anglerfish becomes permanently attached to his large mate. For many animals, the males of the species have no significant role to play other than to contribute to the genetic content of the next generation. The deep sea anglerfish makes very economical use of its males: once a free-swimming male anglerfish finds a female in the murky ocean depths, he attached himself to her, using his sharp teeth. His blood supply becomes fused with hers, and he takes all his food and oxygen from her. His body degenerates until he becomes, in effect, simply a bag of sperm, to be drawn on as required. A female anglerfish may carry several such parasitic males, each one less than a tenth of her size.

(Far right) The paper nautilus or argonaut is a relative of the octopus. The female protects itself and its young with a thin papery shell secreted by special glands. The male is far smaller, only 2.5 to 5 cm long, and lives free like an octopus. But one of its eight arms, the hectocotylus, is quite unique – five times longer than the rest of its body and armed with 50–100 suckers, it lies in a special sac on the male's body. During mating, this arm is charged with sperm and placed in the female's body, where it snaps off as the male retreats. When first discovered inside a female nautilus, the hectocotylus was mistaken for a parasitic worm.

Crab spiders courting. As in many spiders, the male crab spider is much smaller than the female. Mating is a hazardous process for him, as the female is a fierce predator. Fortunately, she is short-sighted, and this gives him a chance to signal to her that he is mate rather than meal. Here he is stroking her to persuade her to mate.

her skin, he takes nourishment from her blood, and he literally loses his senses and his body organs except for his sex organs. In effect, he becomes a parasite on her. Because it is almost impossible for anyone to study life in the deep oceans, little is known about these anglerfish, and many fascinating questions remain to be answered – how does the male find his mate in those inky waters? How does he avoid being eaten by her? And what happens to bachelor anglerfish?

Male parasitism on the female is unusual among animals and occurs when (as with the anglerfish) the chances of the sexes getting together are so slim that once they find each other, they cannot afford to part. Many parasites have this problem. Some have solved it by containing both male and female organs within one body, but many of these hermaphrodites will still mate if they should happen to meet. Among some ticks, the male lies as a parasite on the female; and in one type of roundworm, the tiny male plants himself for life within the female's genital organs. As with any other form of parasitism, smallness is very important for the male in such relationships, so that the female is not debilitated by providing for her inactive husband.

Another marine marriage of opposites occurs with the paper nautiluses, related to the octopus and squid. The female paper nautilus *Argonauta* looks like her larger cousin, the pearly nautilus *(see page 28)* except that her shell is thin and fragile (hence *her* name) and that she can enter and leave

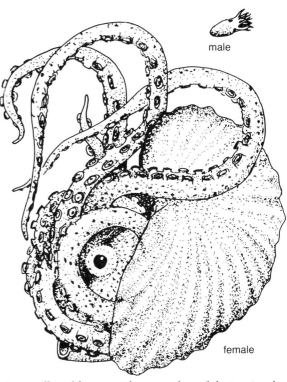

male

female

it at will, unlike any other member of that animal family. She lives in the upper waters of warm seas like the Mediterranean, and may reach 20 cm in length. The male paper nautilus is totally different. He has no shell and looks rather like an octopus except that he is only about 1.5 cm long. The union of this mismatched couple is consummated when the male swims up to the female and, using his third right tentacle which is specially elongated for this very purpose, places a packet of his sperm inside her body. The tentacle breaks off and is left within the female. The male paper nautilus, undaunted, grows a new third right tentacle and then looks round for another female to whom to give it. Until recently naturalists believed that the male and female paper nautiluses were separate species; and that the little tentacle occasionally found inside females was some kind of strange parasitic worm.

Among many of the lower animals, if there is a size difference between the sexes it is usually the female that is larger. The difference is particularly marked among spiders, where the female may be more than three times the size of the male. Female spiders are notorious for their inability to discriminate between a lover and a meal. Rather than cooling their ardour, this simply encourages the males to invent a remarkable variety of ways to distract, trick or immobilize the females so that they can mate and escape. Most male spiders, contrary to popular belief, do live to tell the tale.

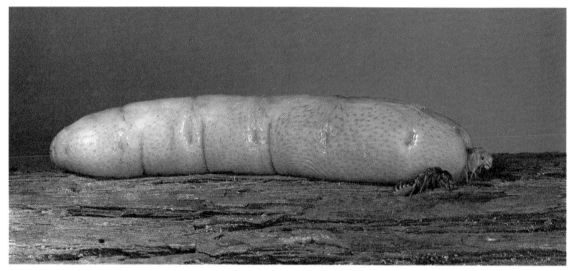

The size of the sexes is complicated among social insects (ants, termites, and honeybees) by the fact that there are males, females, and sterile individuals, within each colony, each of a different size. In the heart of a termite city – mounds and towers of mud, several metres high, containing a maze of passages and chambers – live the king and queen termites. The queen is an enormous bag of eggs. Her soft, pale abdomen may be 10 cm long and 2.5 cm thick – as big as a sausage. The king termite, whose sole purpose in life is to fertilize her eggs, is little more than a centimetre long. Nonetheless he is larger than the other sterile 'castes' in the colony. These include the workers, who are the smallest (they are often immature individuals). A little larger are the minor soldiers; bigger still are the major soldiers whose massive heads and pincer jaws are often larger than the rest of their bodies. Only the king and the queen have the ability to reproduce, although if anything happens to them, some of the juveniles will develop into sexually mature adults. The queen's huge size and her soft body render her virtually immobile and very vulnerable, and she is constantly tended by workers and guarded by soldiers.

A similar society can be found among honeybees, although there are no soldier bees in a hive and quite a few fertile males (drones). The queen bee is not so grotesquely large as her termite counterpart, and there is no such thing as a king bee: the queen's several consorts do not survive their nuptial flight.

Once we enter the realms of the vertebrates we find that if anything, it is the males which tend to be bigger than the females. Most male birds are slightly larger than their mates. An extreme example is the capercaillie, a gamebird of the northern forests: the female, at 60 cm long, is only three-quarters of the size of the male at 85 cm. This general rule is broken by many of the birds of prey, among which the females tend to be larger than the males. The female goshawk, for example, at 60 cm, is 12 cm (20 percent) longer than her mate, and correspondingly heavier. A plausible reason has been suggested for this break with tradition. In a territory where prey is limited, the female bird will tend to take the larger and slower prey animals, while the more agile male can take fast, elusive prey: so between them, like Jack Sprat and his wife, they maintain a balance of resources and avoid

A typical mammal, the male lion is larger, stronger and heavier than his mate. Males will fight for dominance of a pride and hence for the right to mate with the females, and this ensures that only the fittest males pass on their genes to the next generation. This struggle for dominance may even lead the new leader of a pride to kill all the existing cubs in the pride to ensure that only his own progeny survive.

On the breeding beaches, male elephant seals fight for possession of females, and a dominant bull may collect a harem of some forty females. The large bull in the centre is bellowing his claim to the surrounding females. An elephant seal cow may be 4 metres long, but a bull can attain a length of 6.5 metres and weigh 2 or 3 tonnes. They fight by lunging at each other's necks and heads with their canines, but seldom do any serious injury except to calves, which may be accidentally crushed during the fight.

(Far right) Two courting baboons display their somewhat ludicrous sexual attractants – brightly-coloured bottoms. The male (to the right) is much larger than the female.

direct competition. Also, if the male is reluctant to part with his prey at the nest, the female is big enough to bully it from him. She can catch larger, more nutritious prey to feed her young, and she is a more effective guard for them.

Mammalian males, including man, are usually somewhat larger than the females of their kind. This is sometimes associated with masculine strength and the protection of their mates and offspring from predators; and with the rivalry among males for the right to mate with one (or more) females. The larger the male, the more fights he will win and the more females he will impregnate. Among some mammals the largest males will succeed simply by deterrence of size. Females, having nothing to prove by fighting, remain smaller. The female elephant seal, for example, measures up to 3.5 m and 900 kg; the male is far more impressive, reaching 6 m in length and over 3 500 kg in weight, one of the largest of mammals. One dominant elephant seal bull at the breeding grounds may win mating rights to a harem of thirty or forty females. A sad side-effect of their massive size is that during their jealous sparring, the bulls sometimes crush young seal pups to death.

The largest male is often the boss in hierarchical mammal societies, especially among the primates.

A dominant male baboon weighs up to 30 kg, twice as much as the female. The mature male gorilla may stand 170 to 180 cm on two feet, as tall as a man and much heavier – up to 200 kg in the wild, even more in captivity. The female gorilla is comparatively delicate at 150 cm tall and 90 kg in weight. Man, of course, is also a primate, and is not above using his greater size to have things his own way. Many modern sports are ritualized versions of contests between males to decide which is the bigger and better. Were it not for the eventual supremacy of intelligence over bulk, the Goliaths would have subordinated the Davids long ago and the world would be a different place.

Size of young

The size of a young animal is just as important as the size of the adult. It is easy to take for granted the idea that children are smaller than their parents, without considering how much smaller, or why. There are all sorts of factors that govern the size of young animals: the kind of life they are to lead; their metabolic rate and the length of their lifetime; the environment in which they must develop; the type of food they need and its availability; the dangers that they may face, and so on.

As a general rule, small-sized young are produced in large numbers, and large young in smaller numbers. Having large numbers of small young offers better odds of survival to animals which are preyed upon. Predatory species tend to be born larger and in lesser numbers, for being a predator may well involve learning special skills from the parent. This extra investment of time and energy on the part of the parents tends to limit the number of young they produce. Large numbers of small babies probably have no concept of parental care: their mother has laid her eggs, or given birth, in the most likely place for the young to succeed, and left them to make the best of it. Smaller numbers of larger young usually get more attention from their parents. The fewer the young, the slimmer the chances of their generation reaching maturity, and

so the more help they need. If the number of young is small the size of the population cannot change as rapidly as if there are many offspring. Animals are dependent on reproductive ability or on the carrying capacity of their environment.

Simple invertebrates including insects, and lower vertebrates such as fish and amphibians,

This sparrowhawk has just returned to the nest with food for its two youngsters. There is still one unhatched egg in the nest. Birds of prey start incubating as soon as the first egg is laid, so the eggs hatch at different times. This exlains why one youngster is larger than the other. In a bad year, this one would survive at the expense of its brother, rather than both competing equally successfully for food and both dying of starvation.

Frog tadpoles hatching from a mass of frog spawn. Frogs lay large numbers of eggs, then take no further interest in them. During the struggle to reach adulthood, the tadpoles will serve as food for many other pond animals. Since the countryside is not overrun with frogs, it is clear that very few of these tadpoles will survive to become frogs.

Some scorpions give birth to fully formed young. Others carry a parcel of eggs beneath the body. As soon as they are born, the young scorpions climb on to their mother's back, where she carries them for several weeks until they are old enough to fend for themselves.

Large mammals, such as pigs, give birth to well developed young, able to trot to keep up with their mother. Consequently, few young are produced at a time when compared with the young of smaller animals such as frogs and scorpions. This is made up for by more parental care, and by food in the form of milk, giving each youngster a better chance of survival.

generally produce small eggs and young in vast numbers. A female cod, for example, may lay from half a million to nine million eggs in a year, each one smaller than a pinhead in size. The female sturgeon is nearly as prolific, laying up to five million eggs. These vast numbers are designed to counteract vast losses from hostile conditions and predators (although the female sturgeon can hardly be expected to compensate for the predation of man, who calls her eggs 'caviar' and takes them in such quantities that the sturgeon is now an endangered species in some areas). The eggs and young of most insects, and often the adults too, provide food for other animals, so they are laid in hundreds or even thousands, and left to fend for themselves. The dragonfly, on the other hand, lays fewer eggs and positions them carefully, for its larval young is a large and ruthless underwater predator that could soon eat all the prey – and its own brothers and sisters – in an area that was too small to provide for its needs. Predatory spiders and scorpions, too, have relatively few young, but they both provide some parental care. The mother scorpion will carry her young on her back, protected by her deadly stinging tail.

Size is not only relative, but absolute. A large baby has a smaller surface relative to its volume so it keeps warmer, and metabolizes more slowly than a small baby and needs relatively less food. Very small creatures, in general, have many very small young and give them no parental care; medium-sized creatures tend to have medium-sized young in moderate numbers, and to look after them to some

Twin Alaskan brown bears suckle at their mother's breast. Only a few months before they weighed a mere 450 g each. Now they have soft fluffy fur to keep them warm, and they will stay with their mother for about two years.

extent. Large animals mostly have few, large, babies, and give them intensive parental care. Among the many 'medium-sized' animals – including most birds and mammals – the family size is tailored to the number the parent can look after: say a dozen piglets, or five kittens, or four songbird chicks for example. There are plenty of variations on the theme. A chicken can provide for ten or more chicks with ease. A seabird of similar size, like the gannet, must leave its young in a dangerous place and hunt for food which may be scarce, so it cannot look after so many young: it has only one chick at a time.

A problem that usually only occurs among these 'medium-sized' families is that of the small last-born offspring, which must share the home with the larger, stronger first-born. These weaklings, or 'runts' as they are known to farmers, may be

A baby kangaroo finds both warmth and food in its mother's pouch, for this is where her nipples are. There are not many mothers who can run at 25 kmph while carrying their babies! Only 1.5 g at birth, the baby kangaroo does not leave the pouch until about 6 months old. After this it continues to return to the security of the pouch between exploratory sorties outside until it is just too big to get in there. Even then, it may continue to suckle.

A 15-day old American opossum looks more like an embryo than a newborn mammal. As soon as it was born, it made its way unaided through its mother's fur to her pouch and began to suckle. After about three months young opossums leave the pouch but still hitch rides with their mother, clinging to the fur on her back.

elbowed out of the race for life by their own brothers and sisters. In the wild it is common for a percentage of each generation to die of persecution or neglect within the family. It must be the strongest that survive, even if they do so by killing their nursery companions. When the family comprises only one or two young, however, the mother can care for her young individually, and even an undersized infant has a good chance of growing up.

The large animals that bear large young include the bigger mammals – such as sheep or cattle, wild deer and antelopes, lions and other cats, giraffe, rhinoceros, elephant, whales and dolphins (the biggest newborn baby in the world, at 7 metres long and 2 tonnes heavy, is that of the blue whale). Mammal young are usually born well-developed, especially among prey animals that may need to run from danger within the first half-hour of life.

There are, needless to say, some exceptions to these sweeping generalizations. Bears, although among the largest of mammals, have young (usually two at a time) which are relatively tiny, only about 1/350th of their mother's weight. At this guinea-pig size they would have trouble surviving in the wild, not least because their large ratio of surface to volume means that they lose heat rapidly. (Warmth is extremely important to young birds and mammals as they cannot produce enough body heat themselves.) The baby bears can only survive because their mother cuddles them in a warm snug den for the first three months or so of their lives, keeping them constantly warm with her body and nourished with her milk.

The same problem, only more so, confronts the mother kangaroo. She may weigh about 37 kg but she gives birth to a single infant of 1.5 g, about 1/25 000th of her weight. This tiny, helpless mite crawls into its mother's pouch and hangs, literally for dear life onto a nipple. Fed on her milk and kept at her body temperature, it grows at an amazing rate. About six months later the baby kangaroo, known as a joey, is ready to venture out of the pouch. It is now at the same stage of development as the newborn young of other kangaroo-sized animals. For some months yet the joey takes shelter, or hitches a ride, in its mother's pouch, and it will continue to suckle even after a new joey has been born.

Man's nearest relatives, the monkeys and apes, are born at a relatively more advanced stage than human babies, although their weight as a percentage of their mother's is similar. A recent theory suggests that human babies (about one fifteenth of their mother's weight) are born some six months before they 'should' be, if compared to other primates. The reason is that their brains are so large relative to body size that it would be almost impossible for a mother to give birth to a larger baby. The human baby can thus be considered as an embryo outside the womb, rather like – though very unlike in size – the kangaroo joey.

Increase of apparent size

Predatory animals are not concerned to conduct a fair fight with their prey. If it sees a creature that looks as if it might put up up some serious opposition to being eaten, the predator will usually leave it alone and go in search of an easier meal. Some animals have evolved a simple solution to that nasty moment when an enemy stares them in the face: they grow bigger. Small might be beautiful, but there are moments when bigger is better. Being larger than one's opponent tends to impress, and first impressions can be crucial.

Familiar domestic cats and dogs, when alarmed, raise their 'hackles' – the hair on their necks. The cat may even erect the hair all over its body. Anyone who has seen this will know how surprisingly big the cat suddenly looks. It has increased its *apparent* size. If this display fails to deter, the cat may bring other weapons and deterrents into play, but often the enemy will not continue the confrontation. Another creature that raises its ruff is the frilled lizard, a large reptile of northern Australia. If threatened, the lizard opens wide its mouth, and a broad 'collar' of skin around its throat erects like an upside-down umbrella. The frill is also coloured

At the sight of a potential enemy or another male anole approaching his 'territory', the male anole lizard starts bobbing his head up and down and doing 'push-ups' with his front legs while his dewlap – the brightly coloured fold of skin under his chin – is extended in a warning display.

(*Far left*) A yellow cobra threatens an intruder. It raises the front part of its body to give an impression of height, and flattens its neck so that it looks bigger. Then it sways from side to side, probably measuring the distance to its target.

(*Left*) A sage grouse displays on a communal parade ground in the spring. With its tail feathers spread out in a fan and its throat pouches inflated to puff out its breast, it has an impressive appearance. The inflated throat pouches also act as resonators for its loud 'popping' courtship calls.

As the photographer approached this long-eared owl, perched over its prey, the owl fluffed up its feathers and raised its wings in an impressive attempt to frighten him away. Seen from this angle, it looks like a giant among owls.

bright red, unlike the rest of the lizard which is grey-brown. The cobra, too, is famous for raising a 'hood' – the skin at the back of its head – in a threat which, if ignored, may be followed by a deadly strike.

Many birds increase their size when alarmed. If you approach a swan too closely, especially one with a nest nearby, this seemingly serene bird will ruffle up its feathers and half-extend its wings to become a hissing white giant. Owls, too, erect their feathers and thrust their heads forward, increasing their apparent size by two or three times.

Sometimes just standing up straight can have the desired effect. A toad, when approached by a snake, suddenly raises itself tall on its four legs, and the snake finds itself no longer looking its prey in the eye but gazing up at a monster. Other animals that stand up in threat include bears and gorillas. Both are large and powerful animals, but they are not particularly aggressive and are relieved if their enemy turns tail and runs.

Nearly all the animals that use increased size as a threat display have other defences in reserve: the toad's skin emits distasteful slime, and the bear has massive jaws and raking claws. Size for size, however, they have nothing on the pufferfish. This tropical fish, usually about 15 cm long, will swallow water when alarmed until its body is as tautly rounded as a balloon. Quite apart from

surprising a predator, this makes it very difficult to swallow. (Its close relative, the porcupinefish, is covered with spines like a hedgehog. These stick out when the fish inflates and improve the effect.) If the pufferfish's defence is ignored, however, the predator that eats it will have an even worse shock, for the puffer contains tetratodoxin – one of the deadliest poisons known. Japanese gourmets dice with death by eating expensive dishes of raw pufferfish (called *fugu*). Despite meticulous preparation of the fish to remove the poisonous parts, several *fugu* eaters die each year.

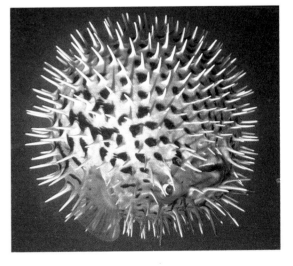

(*Far right*) It is hard to imagine that this spiny pufferfish could look very fish-like! When alarmed, it swallows water to inflate itself, and with its spines erect becomes quite a mouthful for a potential predator to swallow.

The smallest animals

The oceans are vast, and yet, with the aid of a microscope, a bucketful of sea water becomes transformed into a teeming world of bizarre creatures. These are part of the plankton, a name that embraces the thousands of species of microscopic plants and animals that live as drifters in water. Plenty of them live in freshwater ponds and streams, but the richest variety and quantity of planktonic life is found in the oceans. The warmer tropical waters contain relatively little plankton, but the temperate and cold waters up to the polar icecaps may be so full of life, especially in spring and autumn, that they are more like soup than sea. A bucketful of sea water may well contain about 20 000 plants and 150 animals; a cubic metre of sea water averages some 750 000 plants and 5000 animals.

Because many animals feed on the tiny plants, and larger animals eat the animals that feed on the plants, the plants themselves must be the most numerous inhabitants of this micro-world. It was with such organisms, in the primeval oceans, that life started. Even now the plankton contains creatures that are on the hazy borderline between the plant and animal kingdoms. Hardly distinct from plants, they are called the Protozoa, a name derived from Greek which means, appropriately, 'first animals'. They each comprise a single cell of protoplasm – living matter – and some of them are only a few thousandths of a *millimetre* in diameter. Some of them are called flagellates, because of the whip-like tails with which they lash the water to move about. Some of the most diverse and beautiful are the Foraminifera and the Radiolaria. These

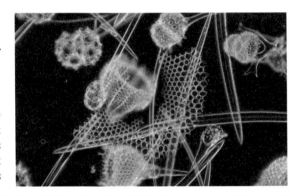

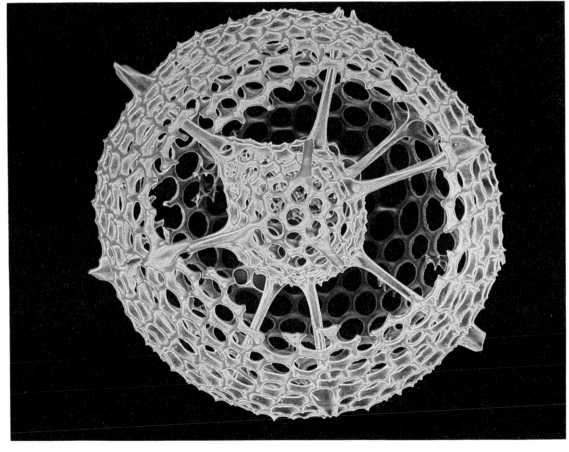

Fossil remains of what has been called the living jewellery of the sea. These are the glassy skeletons of radiolarians, single-celled animals that drift in their millions among the plankton of the oceans. Their remains sink to the ocean floor, and in places make up 30 per cent of the ocean sediments.

creatures have shells or skeletons of glass-like material. Some are rather like snail shells, others are delicate and intricate structures of spheres within spheres. Tiny as they are (twenty or more could sit on a pinhead without touching each other) they are active, and sometimes aggressively predatory. They extend fine 'arms' of their protoplasm through holes in their shells or skeletons to catch and digest their prey. When these creatures die their shells drift down to the sea bed in their millions, year after year, forming a layer of muddy 'ooze'. Gradually over the ages this compresses into rock, which may be moved above the level of the sea by earth upheavals. The white cliffs of Dover are formed of such shells, and geologists can find by studying rocks, often from from the sea, what kind of animal life existed in prehistoric oceans.

Other denizens of the planktonic world include worms: some that never grow longer than a centimetre, others which are the larval (immature)

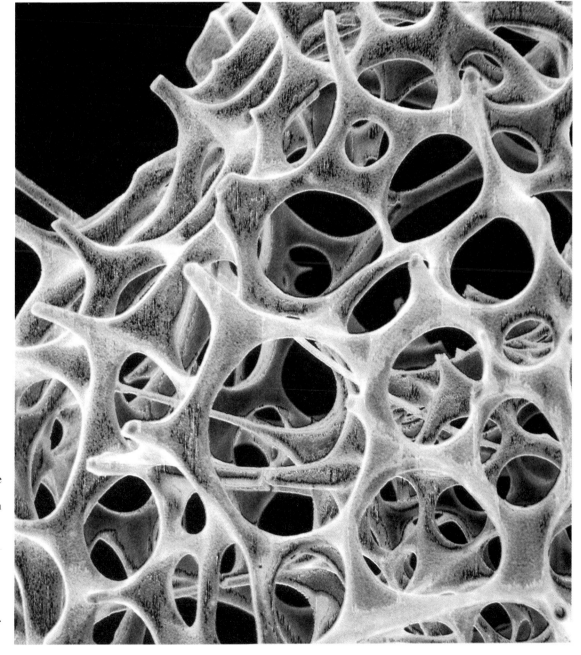

A close-up view of the intricate silica skeleton of a radiolarian. In life, the radiolarian would be a single-celled animal, somewhat like an amoeba but much more complex. The heart of the skeleton encloses the nucleus of the cell, and the rest of the cell material flows around the skeleton, streaming out through the gaps in long fine filaments of sticky cytoplasm which trap other members of the plankton. As well as these delicate protrusions, the skeleton may also be ornamented with spines. The radiolarian's cytoplasm contains a lot of vacuoles, giving the impression of a frothy central sphere through which spines thick and thin project. A group of radiolarians seen under the microscope looks like a cluster of stars in the oceanic universe.

stages of the many types of worms. Many of these larvae, in fact, look quite un-wormlike. Among the planktonic worms are the voracious arrow-worms, with large eyes that can look in several directions at once, including down through their own transparent bodies. Many planktonic animals are totally transparent, an obvious help to them evading capture by predators. Others are beautifully coloured with pinks or vivid blues for no apparent reason. Some, though colourless, reflect light waves in different directions to produce 'interference colours' – dazzling, shifting, rainbow hues.

The molluscs, or snail-like animals, of the plankton include the enchanting pteropods – or, to give them their more evocative common name, sea butterflies. They have minute snail-like shells, but their bodies are modified into 'wings' which flap in the water, keeping the animals buoyant and wafting food towards their mouths.

Although many planktonic animals remain small and pass all of their short lives within this world, others are the larvae of larger creatures and only spend their infancy here. These include the eggs and larvae of many fish species; and larval jellyfish, barnacles, sea urchins, sea anemones, shrimps, prawns, crabs and lobsters. Few of them look anything like the adults they are destined to become. Furthermore, many of them develop through several larval stages, each one looking quite different to the last.

In any sample of marine plankton, up to 90 per cent of the animal life in it will consist of various copepods. These creatures are crustacea, related to the prawns and crabs of the seashores, although few of them grow larger than a millimetre. Like their bigger cousins, they have external skeletons similar to those of insects, and must shed them periodically in order to grow. Copepods look like tiny transparent shrimps. They dart about seeking

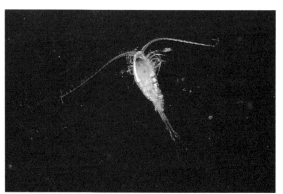

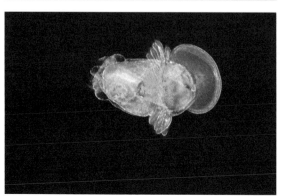

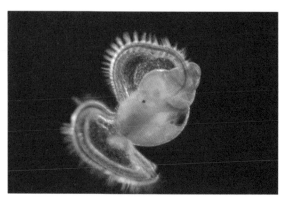

(Above) These sea butterflies live in the mid-deeps of the North Atlantic in large shoals, flapping clumsily through the water. They are sometimes called 'flying angels', but belie their name and prey on other sea butterflies.

(Centre left) A rather unusual deep sea predator – the crustacean *Phronima* preys on salps, barrel-shaped animals with transparent bodies. The female *Phronima* eats out the soft parts of the salp, then lays her eggs in its 'shell', propelling it along with her legs like a floating perambulator.

(Centre right) A copepod with long trailing antennae and a conspicuous red eyespot, probably used to detect changes in light intensity. It swims by means of legs fringed with fine bristles which act like paddles.

(Lower left) The antinotroch larva of a phoronid worm. This larva swims in the plankton by means of ciliated tentacles for several weeks before settling out on the bottom and metamorphosing into a tube-dwelling worm.

(Lower right) The planktonic veliger larva of a marine snail. Veligers have two round head lobes with a fringe of cilia (tiny beating hairs) for swimming. Already equipped with a tiny shell, veligers have to swim quite hard just to keep afloat. They feed on diatoms and other algae in the plankton before sinking to the bottom to metamorphose.

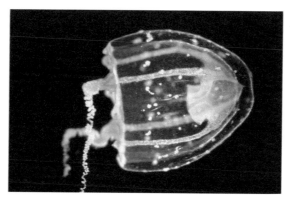

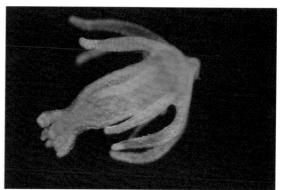

(Above left) This is not a larva, but the reproductive stage of a hydra-like marine animal (a hydroid). The hydroid produces anthomedusae – jellyfish-like forms such as this – which bear the reproductive organs. As they can drift considerable distances in the plankton, they are an important dispersal phase for the animal.

(Above right) A true larva, the 'actinula' larva of another group of hydroids which live near the shore. This larva also helps to disperse the species.

(Centre left) An almost transparent comb jelly swims by means of rows of cilia which beat in waves along its body. Here, some of the cilia are producing interference colours. This is a carnivorous animal, trapping prey on sticky tentacles.

(Centre right) Seen from above, these colourful *Porpita* jellyfish look rather like Catherine wheels. They live at the surface of the ocean, trailing their stinging tentacles in the water below. The central disc is a float to keep the jellyfish at the surface. This is a colonial jellyfish, made up of many distinct individuals which act like a single animal.

(Below) A strange copepod called *Sapphirina* shows dramatic interference colours. Microscopic plates in its surface layers are so arranged as to produce a flash of colour when viewed from certain angles. From other angles, the copepod appears quite transparent.

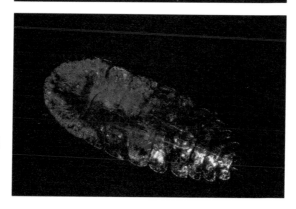

their food – plant plankton, diatoms or algae – or mating. Males may have special hooked arms to attach their sperm packets to the females, and after fertilization the females carry their eggs in two little bags, like shopping baskets, copepods exist in such quantities that they may easily outnumber all the animals that live on land (including birds, insects and spiders).

Also included in the planktonic crustacea are krill, the shrimp-like creatures, 2 to 5 cm long, that swarm in vast multiplications of millions in polar seas. These constitute the main diet of the baleen or 'whalebone' whales, including the largest of all animals, the blue whale. The whale eats by opening its huge mouth and taking a gulp of 'krill soup'. It then squeezes the water out of its baleen plates (a sort of huge sieve) leaving the krill behind to be swallowed. Each baleen whale may swallow several tonnes of krill a day, a tonne of krill constituting an enormous number of individuals; yet still they may occur in such quantities that they colour the sea pink.

Perhaps the strangest thing about planktonic forms is that they come up to the surface of the sea at night and sink down again, sometimes as much as 100 metres, during the day. The reason for this 'vertical migration' is still a mystery. Some scientists think the migration is related to food – the plants come up for sunlight, and the plant-eaters follow. Others suggest that the migration is related to light ; or to water density; or to the movement of ocean currents.

A beautiful characteristic of plankton is their bioluminescence. They contain organs which emit light; this they do at night, mostly in the summer, or when they are disturbed, as by the passage of a boat or a shoal of fish. Then they flash their lights like bright sparks of fire in the water. This phenomenon of 'phosphorescence', which may be seen around our coasts, as well as in distant oceans, is an unforgettable sight.

The largest animals

In contrast to the smallest animals the largest provide striking examples of reaction to evolutionary pressure – mainly from humans. The forms into which they have evolved in relation to their original habitats are discussed elsewhere in this book. But their sizes are nonetheless remarkable and are worth a brief examination.

Modern man's compulsion either to possess or annihilate the biggest of everything has had a sad effect on the largest animals on Earth. All too often they are only seen hanging on the wall of a trophy hunter or chopped ignominiously into petfood and cosmetics. The biggest living creature of all, the blue whale, is almost extinct. The African elephant is endangered; so too are many other land giants including the rhinoceros, the giant tortoise, and the gorilla. Their only crime was to be, in every sense, superlative.

A bull African elephant, standing 3 m at the shoulder and weighing up to 6 tonnes, is the largest of land animals. It is hard to imagine just how big this is unless one sees an elephant, and that is only likely to be in a zoo where its scale and dignity are diminished. Elephants still roam the game parks of Africa, deposed kings of a squandered continent, but their numbers are only a fraction of what they were. Killing for sport, killing for ivory, and killing by land starvation have vanquished the herds of

hundreds that existed even within the lifetimes of most of us.

Even more rare in the wild is the white rhinoceros, slaughtered because – of all stupid superstitions – its horn is believed to be aphrodisiac. The white rhino stands 2 m high at the shoulder and may weigh 3 tonnes. Perhaps more than any other animal it recalls the age of the dinosaurs,

Although it looks menacing, a white rhino is quite a docile animal unless provoked. But a mother rhino with a calf must be treated with great caution, as she is extremely protective and will charge any intruder and attempt to impale him on her front horn, which may be over a metre long.

Millions of years ago, ancestors of the elephants roamed most of the earth. The African elephant is the largest living land animal. It has massive legs, and its feet may be up to half a metre in diameter. The elephant requires such large quantities of plant food that it needs to roam vast areas in search of it. It is an intelligent animal with an excellent long-term memory, probably needed to memorize trails and good feeding places for future reference.

The gorilla is now an endangered species, mainly because of the cutting down of the tropical forests which are its home. Despite its great strength, it never kills to eat, but lives on fruit and vegetables. Young gorillas climb in the trees, but adult males, up to 180 kg in weight, are too heavy for acrobatics, and live on the ground. Gorillas live in family groups or small tribes, and keep on the move, as they need to cover a large area to get enough food. At night they sleep in trees or in nests of leaves and twigs on the ground. It is the need for considerable areas of undisturbed natural habitats that is the chief cause of the decline of the gorilla population, as it is of the elephants.

those vast and apparently terrifying prehistoric creatures. In fact the largest land mammal ever to live, *Baluchitherium*, was a cousin of the modern rhino. This great ungainly animal, 5.5 m at the shoulder, 11 m long and about 20 tonnes in weight, was a browser of plants.

Apart from the elephant and the rhino, Africa claims many other animal giants. The giraffe, tallest of animals at 6 to 7 m, gallops with slow-motion grace across the savannahs. The Nile crocodile is among the largest reptiles at 4 m long. The ostrich, largest living bird, stands 2.5 m tall; the hippopotamus may reach 4.5 m in length and more than 2 tonnes in weight. The reticulated python, 6 m or longer, is the world's longest snake. The gorilla, largest of the primates, is one of man's closest relatives, and for all its size is amiable in demeanour. It seems that man may just about have discovered the true relationship between himself and his ancestors, the great apes, by the time he has rendered them extinct. Africa was the cradle of man, but he has transformed it into the coffin of the gentle giants.

Unlike most of the flightless birds in the world today, the ostrich, denizen of the African plains, evolved in an environment with no shortage of mammalian predators. Its height allows it to spot predators some distance away, and its powerful legs will transport it at up to 64 kmph out of harm's way. Even if cornered, these legs and feet can deliver a formidable kick, enough to deter any carnivore.

The reticulated python holds the record for being the longest snake in the world, at exactly 10 metres. It is also the only snake reliably documented to have swallowed a human being: a 14-year-old boy from Talaud Island in Malaysia. Pythons usually kill by constriction, coiling their bodies around their victim and squeezing until the prey is unable to breathe.

In ancient Egypt, not only the Pharaohs were mummified, but also the Nile crocodiles. The largest Nile crocodile recorded was 6.5 metres long and must have weighed about 1050 kg. The Nile crocodile has a bad reputation, and probably kills nearly 1000 people a year – some 'revenge' for the crocodile lives lost to the skin trade.

The giraffe has the distinction of being the tallest living animal, a bull giraffe averaging about 5 metres from hoof to horn. The tallest recorded giraffe was 5.87 metres tall – 1.21 metres taller than a London double-decker bus. In the African savannah the giraffe uses its height to browse trees at a level inaccessible to other mammals, so there is little competition for its food source.

External skeletons

The skeleton of a vertebrate provides a framework around which the organs and muscles of the body are hung. But many animals display their skeletons for all to see: their surfaces and their skeletons are one and the same thing. These animals with external skeletons include the insects, the largest group of species on earth; arachnids – a group that includes spiders and scorpions; myriapods – the centipedes and their relatives; and crustaceans – the prawns, crabs, lobsters and their kind. These groups are known collectively as arthropods, or 'animals with jointed legs'.

The external skeleton of an arthropod is known as its cuticle (see diagram). It consists of several layers of material, produced by a skin-like surface all over the animal, called the epidermis. This secretes a mixture of proteins and a tough fibrous substance called chitin, which combine and harden to form an extremely tough, plastic-like compound called sclerotin. The inner layer of

cuticle, the endocuticle is soft, but closer to the outer surface the layer called exocuticle may be tanned and rigid. In most insects these two cuticle layers are less than 1/5th mm thick. Forming an even thinner (1/500th mm) cover over them is the outermost layer, the epicuticle. This contains an extremely fine, waxy layer for waterproofing, and a final top-coat of varnish-like 'cement'.

An insect needs a rigid cuticle as it has no other means of support or protection for its soft body organs. If the cuticle were uniformly stiff all over, however, the insect could not move; so in the region of joints the cuticle is softer and more flexible. The result is a system of plates linked by moving joints, like the suits of armour worn by medieval knights. The waterproof wax layer is designed not so much to keep moisture out as to keep it in. With such a large surface relative to body size an insect risks drying out by evaporation. Only a few primitive arthropods have no waterproof

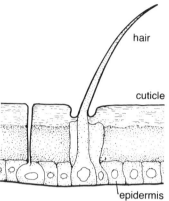

An insect hair is a sensory structure that grows through the cuticle from the epidermis, where it is connected to the insect's nervous system.

(*Above left*) The male stag beetle's exoskeleton is used to form a formidable pair of antler-like jaws. When accosted, he rears up and opens his jaws wide to threaten the enemy. However, if his bluff is called, the intruder will find that his bite is not so painful as that of many smaller beetles.

(*Above right*) The black widow spider's feet possess small claws that enable her to hang beneath her web. The abdomen has a bright patch of red pigment to warn would-be predators of her poisonous bite.

(*Lower left*) The woodlouse has an exoskeleton very clearly divided into segments, with softer bands of cuticle between. This enables it to bend its body round narrow crevices, and even to roll into a ball if threatened.

(*Lower right*) A grapsid crab has a very tough exoskeleton, able to withstand buffeting from the waves. Its powerful claws are used as pincers to seize prey or to grapple with predators. Surprisingly, if this crab is itself grabbed by a predator, it may shed its claw to escape, then grow a new one at the next moult. While the predator attempts to remove the claw which still grips it, the crab makes its getaway.

This close-up of a locust shows the many forms the exoskeleton can take: finely jointed antennae, coarsely jointed legs, claws for gripping the leaf, mouthparts for biting, and associated appendages for guiding the leaf into the mouthparts, a tough plate over the thorax, but fine membranous wings, and through all this a camouflaging pattern of spots.

A newly emerged locust hangs from a cactus stem while its wings are pumped up with blood. Once fully expanded, they will harden and the blood will be withdrawn. Locusts mature in stages: at each moult the newly emerged insect will have more highly developed wings than the previous moult. For this locust, there will be no more moults – it has reached its full size and is now adult.

layer, and can absorb oxygen through the cuticle from the moisture around them. Most of the arthropods must compensate for their impermeable coat by having tiny holes, or spiracles, through its surface to admit oxygen in the air into a network of inner tubes or tracheae. Where air can get in, water can possibly get out, so spiracles often contain little valves to minimize water loss.

A final component of the armour of some arthropods is a substance called resilin, very like pure rubber and the most elastic natural material. It is hard when dry, but usually contains 50 to 60 per cent of water. Little pads of resilin are situated in places where strong sudden movement is required, as at the base of insect wings. When the wing is raised the resilin springs back to shape, pulling the wing down without using body energy or incurring friction drag. At the base of its hind legs, a flea has resilin pads which work like a catapult to shoot the insect into the air – as much as two hundred times its own length.

Once the cuticle has hardened it cannot expand or change its shape. So, as it grows larger, an arthropod must periodically shed its coat of armour and replace it with a new and larger version. In many cases the animal changes its shape and characteristics at the same time; this is the strange

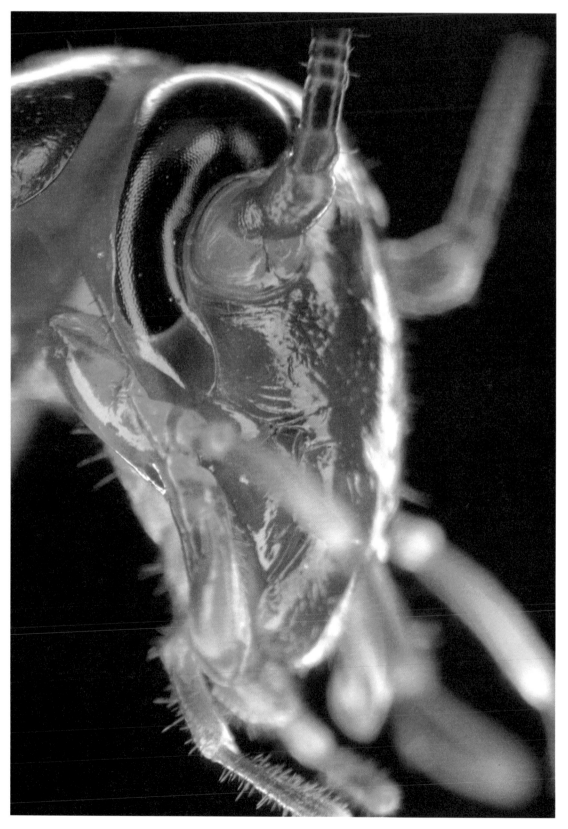

The head of a European house cricket shows the large compound eyes and the jointed base of its long antennae. It is omnivorous, coming out to feed on household scraps in the evening and hiding in dark crevices during the day. Much of its food will be detected by smell, hence the long antennae which are armed with tiny smell detectors. The sensory hairs are much more numerous on the palps, which are used once it has found its food. Note how the cuticle is thickened in plate-like sections, with bands of thinner cuticle in between. This allows flexibility of movement.

The head of the brown aeshna dragonfly shows an enormous pair of compound eyes which almost encircle the head. The dragonfly is a fierce hunter, catching insects on the wing, and for this it needs very accurate vision.

and miraculous process known as metamorphosis. The most familiar example is the caterpillar that changes into a pupa that changes into a butterfly. Before becoming a pupa, however, the caterpillar moults its softer cuticle several times without changing its basic shape, to allow for growth. Some adult arthropods, such as crabs, shed their cuticle and emerge with the same pattern in a larger size.

An insect grows inside its cuticle before moulting. The new cuticle, soft and wrinkled, is formed underneath the old one. Gradually most of the old cuticle is absorbed back into the body leaving the dry, dead, hard outer layer. When the skeleton is ready to burst from the pressure of new body tissue, chemical signals are sent to tell the old cuticle to split, usually along the back. The insect swallows air and increases its blood pressure to expand the

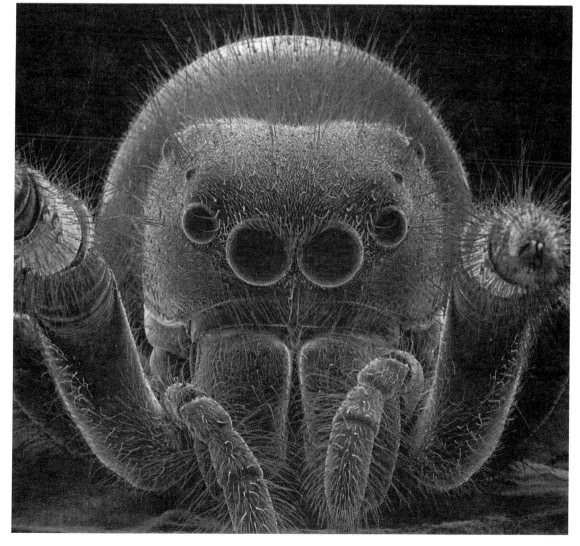

The scanning electron microscope reveals the fine detail of a jumping spider's head, magnified 35 times. Jumping spiders are said to have the best vision of all invertebrates. They catch their prey by ambush, and the large eyes placed close together in the front of the head give accurate stereoscopic vision for judging distance when pouncing. The remainder of its eight eyes give it almost all-round vision, so it can spot prey moving anywhere near it.

(*Far left*) It might seem strange that a prawn should sport such a brilliant, conspicuous colour, and even stranger that at the depths at which this prawn lives (500 metres) it appears black. In the ocean depths, only blue light penetrates, so there is no red light to be reflected as colour – the apparently gaudy coat is a perfect camouflage.

(*Left*) A desert scorpion fluoresces when viewed by ultra-violet light. No one knows what benefit this is to the scorpion, but it is very useful to scientists hunting scorpions in the desert at night!

soft new cuticle as much as possible before it hardens on exposure to the air. Meanwhile it emerges from its old skin and for a while – which may range from minutes to hours, depending on the species – it is soft, incapable of movement, and very vulnerable. Finally more chemicals are produced to 'tan' the cuticle in its new fully-expanded shape.

The sclerotin of cuticle is a remarkable material, with even more strength and versatility than its man-made counterpart, plastic. All external structures of arthropod bodies are formed from it: eyes, antennae, jaws, legs, wings, claws, spines, scales, etc. The jaws of some insects can cut through wood and even metal.

On most insects the cuticle bears hair. These serve as sensory organs, touching, tasting, or detecting wind direction. Each hair passes through the cuticle from the epidermis, where it is connected to the nervous system. On many flying insects like bees there is a thick layer of hair over the body, especially over the flight motor in the thorax, to keep it warm and functioning well. Cuticle coats come in many colours, from the transparent wings of a dragonfly to its brilliant blue body; from the basic brown of many insects to the shimmering iridescence of others. It is an excellent medium for camouflage. The shape and colour of the cuticle may mimic, to an incredibly accurate degree, a leaf, twig or flower, or even a bird dropping.

Cuticle coats can grow thick, chalky and heavy, as in crabs and lobsters. The largest of these armour-plated skeletons is that of the Japanese spider crab, whose leg span may reach 3 m. Such size creates difficulties, for to support a large body the cuticle must become thicker and heavier,

reaching a point at which it becomes too heavy for the animal to carry easily. The largest external skeletons belong to sea creatures that can absorb oxygen from the water through gills. On land the arthropods must obtain oxygen from the air, through tracheal tubes along which the air diffuses slowly so there is a limit to the distance along which this form of respiration can be effective. Overall size is therefore restricted. Moulting the cuticle also becomes a greater problem as the animal's size increases. During the moult it is more vulnerable for longer periods, and there is nothing to support its body organs. The external skeleton is very efficient, but only to a limited size.

An extravagance of branching spines decorates this Venezuelan caterpillar's body. These spines are used for defence, and contain chemicals which cause great skin irritation if the caterpillar is touched.

Like a piece of leaf rotting at the edges, a membracid bug remains perfectly still to complete its protective deception. This family of bugs contains many great mimics which resemble leaves, thorns and other objects in their environment. Yet there is an underlying design which distinguishes them as bugs, distinct from all other groups of insects.

Another bug, the jewel bug from tropical Australia, quite different from the one above, conspicuous rather than camouflaged. Both these bugs feed by sucking up plant juices, piercing leaves with their sharp mouthparts. This jewel bug is actually feeding: its long stylets hang down between its front legs and penetrate the main leaf vein.

Skin

Skin is definitive: it defines the shape of most creatures. Primarily the skin must contain and protect the living cells that constitute the animal. In wet conditions it must provide waterproofing; in dry, it must retain the body's moisture. It must monitor the difference between the external and internal temperatures and help the animal to achieve a comfortable balance. It must alert the animal to the proximity of other objects, and if any foreign body attempts to invade, the skin must help to identify, locate and repel that foreign body. If it is damaged, it must repair itself as quickly and neatly as possible, to prevent the entry of foreign organisms and the loss of living cells.

The skin fulfils all these tasks as well as many more. It may mould weapons and build defensive structures. It may adopt a useful colour, or change its colour to conform to the environment. It may manufacture products that help the animal to move or function better, others that encourage or

A male frigatebird or Man-o'-war bird has selected a suitable nest site and is advertising for a mate by inflating its crimson throat pouch. As soon as the first egg is laid, the pouch will be deflated.

The red-faced uakari has a quite naked face with bright red skin, which glows when it is excited. Uakaris have short tails with bushy ends, and are the only short-tailed monkeys in the New World. They are confined to a very small part of the Amazon forest, staying high in the trees and feeding on fruit and insects.

deter different organisms that come into contact. It may grow thicker or thinner in different regions according to local needs, or temporarily adjust its shape to communicate a message from the mind.

With all these tasks to perform almost incessantly during the lifetime of an animal, it is not surprising that the skin is complex. In most vertebrates the skin consists of two principal layers (*diagram a*): the outer layer or epidermis, and the inner layer or dermis. The epidermis is itself layered, and the outermost layer is dead. As the cells of the inner layers of the epidermis are pushed out towards the surface, they die and become converted to the protein keratin. Keratin is tough, adaptable, flexible, resistant to water, and provides a good protective covering for the rest of the body. These qualities also make it an ideal material for the moulding of claws, nails and hooves, all derived from skin. As it is constantly replaced from underneath, the epidermis flakes away or is rubbed off at the surface. In some places, especially where friction occurs, it is thicker: tough dead skin can be cut from the fingers or feet without incurring pain, for the epidermis contains no nerves. The epidermis

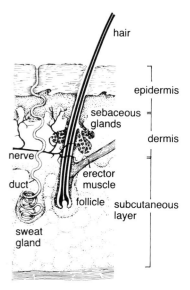

(*a*) A section through the skin showing a sweat gland with its coiled duct (*left*), and a hair in its follicle. To the right of the hair is the diagonal muscle used to erect the hair, and above this a group of sebaceous glands open into the follicle. A branching nerve entering from the left has endings around the hair, to detect any strain on the hair, such as if it is pulled.

is also folded and wrinkled in characteristic patterns – sometimes so different that, for example, of all the hundreds of millions of human beings in the world, no two individuals have identical fingerprints.

The dermis is attached with many little projections that extend into the underneath of the epidermis. It contains blood vessels, lymph vessels, nerve endings, muscle fibres, hair follicles with hairs and sebaceous glands, sweat glands, mucus glands, poison glands, oil glands and scent glands. Unlike the epidermis, the dermis is very much alive.

Blood vessels bring nourishment to the skin, and help to control body temperature. When the body is warm the blood flows more copiously towards the skin, giving off heat to the cooler air outside; it flows even more strongly when the body is being exerted, so someone taking exercise on a warm day will be 'flushed' from the extra blood passing through the skin for cooling. Conversely, in cold weather small valves may close to restrict the blood supply to the skin and hence conserve heat within the body. When this happens, our fingers turn white. Skin will also redden, due to the sudden dilation of blood vessels, for psychological reasons, the most familiar being the phenomenon of blushing with embarrassment. It is not clear what function this ever

served in mankind, but it occurs in few other animals – although the octopus may suddenly turn red if provoked. The cooling function of blood in the skin is heightened in some animals by the use of very large, thin areas of skin – usually the ears – where the maximum contact with air is possible *(see page 170)*.

Nerve endings are scattered liberally over the skin, making it the largest of the body's sense organs. There are extra nerve endings attached to the roots of hairs and bristles *(see page 82)* so that these are particularly sensitive too. It has been calculated that a pressure of $1/28\,000\,g$ at the end of a hair 9.5 mm long is enough to trigger a response from the nerves. Human fingertips, probably the most sensitive skin areas in the animal world, can distinguish between a smooth glass surface and one bearing grooves only $63\,000\,mm$ deep.

The muscle fibres in the skin and in the muscle layers just underneath it can make the skin twitch – often a reflex response, as when a fly lands on a cow's neck. The muscles attached to hairs can erect them, either in threat or for temperature control. In primates and especially man, muscles move the skin to produce a great variety of facial expressions which are vital for facial communication.

A close-up of the skin on a human finger. The flexibility of the skin, together with the fine ridges, enable the fingers to grip objects and manipulate them. The pattern of dermal ridges is unique to each individual human being, and even to each finger, and is used by criminologists to identify suspects. This part of the skin is very well supplied with touch sensors, which enables people to identify very fine textures and structures, like, for example, the braille characters 'read' by the blind.

The bright colours of this poison arrow frog are a warning to would-be predators that it is poisonous – its skin contains many poison glands. South American Indians scrape the droplets of liquid from its skin, allow it to ferment, then dip their hunting darts in it. The venom is not lethal to humans, but can kill small birds and mammals.

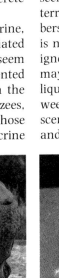

(*Right*) A cocker spaniel pants on a hot day, its tongue hanging out of its open mouth. The evaporation of water from the moist tongue cools the blood in the fine vessels supplying it, helping to keep the body temperature down.

(*Far right*) The striped coat of the skunk is a warning to intruders that it can be offensive. If threatened, it will turn its back on the intruder and squirt a nauseous smelling fluid from its anal sacs, often with surprisingly good aim.

Sweat glands play an extremely important part in temperature control. Shaped like a tube, knotted at the bottom and opening out of the epidermis at a 'pore', sweat glands secrete a colourless liquid which evaporates on the surface of the skin, removing excess heat. Many animals, however, have few or no sweat glands. Birds, reptiles amphibians and fishes have none. Animals in cold climates usually do not need them, and have very few. Whales and dolphins have none, being kept cool by the sea water around them. A beached whale may die simply because it gets too hot; it has no way to rid itself of the excess heat produced by its body metabolism, so eventually it bakes in its own skin. Dogs have few sweat glands, and cool the blood by panting: evaporating saliva from the tongue hanging out in the air. Horses, on the other hand, have plenty of sweat glands which secrete copiously when the horse is exercised.

There are two kinds of sweat glands: apocrine, associated with hairy skin, and eccrine, associated with smooth, hairless skin. Apocrine glands seem to be concerned mainly with producing scented secretions, and are progressively replaced in the more advanced mammals – gorillas, chimpanzees, and especially man – with eccrine glands, whose secretion dilutes and spreads that of the apocrine

glands. It is the few remaining concentrations of apocrine glands on human beings, in the armpits and groin, which are responsible for human body odour. Human beings generally do all they can to remove this scent and replace it with spicy or flowery perfumes. In the animal world, body odour is extremely important, disseminating information about an animal's gender and sexual condition, for example.

Derived from sweat and sebaceous glands are the specific scent glands which are a distinctive feature of many mammals. They may occur almost anywhere on the body: under the tail, as in the rabbit; on the feet, as in the rhinoceros; on the belly, as in the musk deer; on the back, as in the peccary; on the neck, as in some marsupials; even on the forearm, as in some monkeys. The odoriferous secretions of these glands are used to mark territory, advertise sexual condition, identify members of a clan, or even to repel predators. The skunk is notorious in this respect, and any animal that ignores its black-and-white warning coloration may get hit, even at a distance, by a jet of pungent liquid which will cause distress for days or even weeks. Some animals are hunted by man for their scent glands, especially the mongoose-like civets and the Himalayan musk deer: their scent glands,

(*Right*) The Indian rhinoceros has an extremely thick skin with a pattern of folds that gives the impression of a suit of armour.

(*Far right*) The hippo's skin is so heavily lined with fat that it weighs almost half a tonne. Its blue-grey colour camouflages the hippo while it is lying in the water.

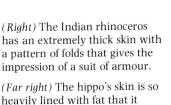

civet and musk, are important in the perfume industry. Civets are to a certain extent bred for this purpose, but musk deer are hunted mercilessly and are now very rare.

The oil gland at the base of a bird's tail, secretes a liquid which the bird spreads over its feathers to condition and waterproof them. The gland is most developed in water birds and is important to their buoyancy. Fish and amphibians have a great many mucus glands. In amphibians the mucus is extremely important to keep the animal's skin constantly moist, which is necessary for respiration. Amphibians take in oxygen through their skins as well as their lungs: up to 20 per cent of the total intake in a salamander, for example, is through its skin. The mucus also keeps their thin skin supple and protects it by creating a very slippery surface which predators find hard to grasp.

Yet another type of skin gland is the poison gland which is prevalent among amphibians, including the common toad. The 'warts' on its skin are actually bunches of poison glands. The poison they give off is not very strong; it might irritate the skin of a person picking up the toad, or leave a dog wiping its mouth in disgust. A similar but stronger poison is found in the skin of the European fire salamander, and it may have gained its name from

the burning pain it causes if handled. The strongest skin poison – in fact one of the deadliest poisons known to man – is found in the skin of the 'arrow-poison' frogs of South America. These small frogs are often brilliantly coloured as a warning to predators. The native indians catch the frogs and impale them on sticks over a fire, which causes the skin to emit the poison mucus. The indians tip their arrows with the poison and use it in hunting and even warfare. The most lethal frogs are a few *Phyllobates* species: named batrachotoxin, their poison is 250 times stronger than strychnine, and acts on the nervous system.

Skin may look and feel very different on different

Another brightly coloured poison arrow frog. Frog skin is always moist: unlike the skin of reptiles, birds and mammals, it is very porous, and is used as an extra respiratory surface. This has the disadvantage that water can also diffuse across the skin and evaporate, so frogs need to live in a moist environment.

A male spring peeper is singing in the spring. During the mating season all the spring peepers of a particular area converge on a pond and 'sing', their inflatable vocal pouches giving extra resonance to their calls. Female frogs do not have vocal pouches. The chorus of male frogs attracts females to the pond, and is said to sound like sleigh bells in the distance.

Newborn harvest mice are quite naked, developing fur only after a few days. For the first week their eyes and ears remain closed. To help keep them warm until their fur develops, their mother has to build a snug nest of corn leaves, where they lie huddled together for warmth. If they were more developed at birth, they would also be larger, so harvest mouse litters would have to be smaller. Such helpless young are produced only by animals that will provide parental care and protection against predators.

creatures. It may be thick – 6 cm thick on the neck of a walrus, for example – where it protects and insulates, especially when lined with a layer of fat or 'blubber'. The skin is an elastic and light fabric for all kinds of delicate body structures, such as bat wings. Bats are the only mammals that fly. Their wings consist of two layers of skin containing blood vessels and muscle fibres, which shrink at rest and are stretched in flight like an umbrella on its spokes. Membranes of skin are also used by other animals to achieve, if not real flight, at least gliding. The flying squirrels and gliding possums have a membrane of skin extending between the wrist and ankle on each side of the body. When the animal launches itself from a high branch it spreads its limbs wide apart and the taut membranes act as a parachute: the great gliding possum can make leaps covering 100 metres in this way. 'Flying' frogs have similar enlarged membranes between their long toes which they use in gliding leaps from tree to tree. Many aquatic animals use membranes on their feet as water paddles (see page 180).

The skin also fashions fine bags and pouches. Kangaroos, wallabies and other marsupials have skin pouches to keep their developing young secure, warm and concealed. The pelican uses the unique pouch of skin in its lower beak as a fish-trap; it flies low over the water and, on seeing a fish, dips the lower beak into the water to scoop it up. The pelican does not store food in its beak as is often supposed, but swallows it immediately, regurgitating it later if necessary to feed its young. Skin can also be used to make 'music', as demonstrated by the nocturnal chorus of frogs in warm climates. The frogs gulp air until their throat skin is distended like a balloon, then exhale it sharply to produce a

wide variety of sounds designed to enamour the females of their kind.

A final quality of skin is its colour. Where it is very thin it is sometimes shaded pink by blood vessels showing through it. It may also take on various shades of reddish-brown, yellow-brown or even black, due to the action of the pigment melanin. As every seaside sun-worshipper knows, melanin helps to protect the skin from sunlight, but in man it is the quality and not the quantity of melanin cells that counts. Most mammals and birds are rather dull as far as skin coloration goes, and the exceptions are all the more astonishing – the red nose and blue cheeks of the mandrill baboon, or the turquoise neck of the cassowary bird, are splendidly impressive. In fishes, amphibians and reptiles there are no inhibitions about colour, and all the shades of the rainbow may be found in their skins. Many fish skins contain mirrors made up of stacks of thin plates of minerals spaced exactly 0.5 µm apart – these act as interference mirrors.

The mandrill is said to be one of the most vicious of the primates, guarding his females jealously and punishing infidelity with death. The male's brightly coloured face may serve as a warning to other males, or as an indicator to females of his status or condition.

Hair

Hair is one of the special characteristics of mammals. Most mammals are covered with fur, hairs, bristles or a combination of these. The exceptions, almost or completely hairless, include whales and dolphins, elephants and rhinoceres, pigs and tapirs, and to a certain extent, ourselves; the hair on most humans is limited and localized. Hair is very useful, contributing to the success of mammals by providing an easy, adaptable and effective method of temperature control.

Hair is dead. By the time each hair reaches the surface of the skin it has keratinized – hardened and died. The only living part is the root inside the skin. Each single hair starts its brief life at the bottom of a little growing-pit, called a follicle (diagram a). At the base of the follicle the hair is attached by its clubbed base to a protrusion of the follicle called a papilla, which contains blood vessels that nourish the hair. Because this growing root is alive, pulling a hair out of the skin causes slight pain, while cutting hair is not painful.

Most follicles have their own tiny muscles which can be contracted, pulling the hair into a vertical position. This is what happens when you have a fright: your hair 'stands on end', and your skin is suddenly covered in goose-bumps. On a furry animal the effect can be a sudden increase of size, and raising the hair on the neck or all over the body is a common signal of alarm or threat. Most of the time, however, the individual hairs lie at an angle, close to the skin surface. Also attached to the follicle is a sebaceous gland, exuding a greasy wax called sebum which coats the hair with a waterproofing, lubricating, and disinfecting layer as it grows up through the follicle. The sebum exuded by sheep is particularly rich and is used, as lanolin, in cosmetics. Sheep-shearers traditionally have very smooth, supple skin on their hands for this reason. Follicles may be evenly spaced in the skin, but in many animals such as camels, sheep, deer and wild goats, the hair follicles are arranged in groups in the skin.

Each hair has different parts (diagram b). The main fabric of the hair, forming most of its cross-section, is the cortex. Inside this there is sometimes a central core called the medulla, hollow or meshed, which occurs in thick hairs. On the outside of the cortex is the thin layer called cuticle, which is formed of flattened cells that more or less over-lap each other; because of this they are called scales. It is not apparent to the naked eye that hair has a scaly surface, but sometimes a hair drawn between

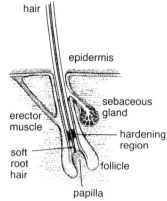

(a) Section through skin showing a hair follicle.

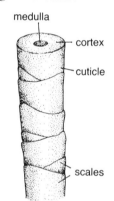

(b) An individual strand of hair.

(Far left) On a rocky pinnacle above the tree-line, and even above the clouds, a Rocky Mountain goat surveys his domain. One of the few mammals to keep a white coat all year round, his long hair has a thick undercoat of fluffy wool. The long outer coat protects the goat from icy winds, rain and snow, and is shed in summer.

(Left) Three lazy cheetahs look very conspicuous stretched out on the grass, but against the sandy colour of the semi-arid regions where they normally live, their body colour blends with the environment. The spots break up the body form so that the overall shape is difficult to discern.

Members of the cat family have particularly large whiskers around their nose and mouth. These whiskers are connected to special groups of nerve cells and are very sensitive to touch, which is important when hunting by night.

An encounter between a porcupine and a striped skunk. Both animals have some very long striped bristles, which they use in defence. Usually the porcupine's quills lie flat along its body, but here it is erecting them to make it appear bigger and fiercer. The skunk has erected its magnificent bristly tail, a warning of its main weapon – the foul-smelling irritating liquid it can squirt from anal sacs just under its tail.

the fingertips feels rougher passed in one direction than the other. Under a microscope the scales of a hair are clearly visible and it can be seen that each animal species, even each individual, has its own hair scale pattern, as unique as a fingerprint.

There are two major types of hair: guard hairs, which are long, thick and tough, and under-hairs, which are usually fine, soft and silky. Fur is typically seen in rodents and consists of under-hairs growing close together. The wool of sheep is also mainly composed of under-hairs, very densely packed in some breeds. At the opposite end of the scale is down, the short, fine, sparse under-hairs that cover the heads of babies and bald men, for instance.

Guard hairs may be flattened in shape; this adaptation is most marked among aquatic mammals such as seals, so it is evidently a waterproofing device. Guard hairs may also become thick, stiff bristles or whiskers. In many animals bristles have a sensory function: their roots are well supplied with nerve endings which provide immediate information to the brain if the bristle is touched. The whiskers of a cat, for example, are extremely sensitive and help it to find its way – and its prey – in the dark. Finally, guard hairs may become even longer and more rigid than bristles, and end up as spines. Some animals have just a few spines. Some are covered with them, like the familiar hedgehog and the even more daunting porcupine. Spines are, obviously, defensive; and they are so effective in that respect that one sometimes wonders how hedgehogs or porcupines ever manage to get close enough to each other to mate or have babies. In fact the spines of a baby hedgehog are soft at birth and do not become stiff for a couple of days – otherwise the mother would have a hard time giving birth. The quills of the giant porcupine are not only up to 38 cm long, they also carry barbed tips to ensure that once embedded in an enemy they are not easily removed. There are accounts of tigers being permanently lamed as a result of infection from wounds caused by these irremovable spines.

Hair grows at different rates. There is seasonal growth: many mammals grow a light coat in summer and a denser coat in winter; this winter fur has hairs with a larger cortex, so trapping more air. There is even a slight seasonal change in the growth of human hair. It has been noted that in the northern hemisphere the autumnal equinox is the signal for many different mammals to start growing their winter pelage. This often involves more than a change of texture for animals that live in the colder regions of the world. They may shed a coloured summer coat for a white winter one which blends with the snow. The common stoat, in the north of its range, sheds a summer coat of red-brown with a white belly, and becomes white with black tail-tip. It is then known as an ermine, and its winter fur being used traditionally to border royal robes. At any time of year, health is important to hair. A healthy, well-fed animal has a more even and glossy coat than a sick or hungry one.

Most mammals have coats coloured from black through grey to white, or brown through red and yellow, or a mixture of those shades. The colour is formed by the pigment melanin at the hair root, colouring the cortex and often the cuticle of the hair. The colour may be affected by the health of the animal, the season, the strength of the sunlight, and other factors. In several animal species as well as man, hair gradually loses its colour and turns grey with age.

Mammals are warm-blooded and hair is a most adaptable method of helping to maintain an even body temperature. Generally a dense coat of under-hairs, as in the wool of a sheep, is particularly effective in temperature control, because hundreds of tiny air pockets become trapped among the hairs and make an insulating layer between animal and climate. Sheep with thick wool, such as the merinos of Australia, can stay warm in freezing weather and, conversely, stay cool in the heat of summer. In both cases the difference between the temperature at the skin and on the wool surface (a distance of 8 cm) may be 40°C or more. In animals with less thick coats, simply erecting the hair increases the resistance to cold. A cat's fur is sleekly flat on a sunny day, fluffed up in the chilly evening. Naturally, animals that live in polar regions have the warmest coats of all. The reindeer's coat combines long, water-repelling guard hairs with an extremely dense underfur, deep-piled like a carpet. The musk ox has shaggy outer hair and a thick silky undercoat in which it can withstand any blast or blizzard, and polar bears live comfortably on ice. The case of the arctic fox is interesting as it does not have to increase its metabolic rate until the temperature falls below –40°C.

A white-tailed deer fawn waits motionless in a patch of sunlight on the forest floor, relying on camouflage to protect it from predators. Its general brown colour blends well with the leaf litter around it, and its spotted back and sides resemble the dappling of sunlight in the forest, as well as breaking up its outline when viewed from a distance.

The coat that kills! A leopard skin is so prized by hunters and the fashion trade that its rightful owner is on the decline. Originally designed to conceal it, the beautiful pelt now attracts unwelcome attention. The bronze colouring is much darker on its head and back than on its underparts, which are almost white. This actually makes the animal less conspicuous: normally objects standing in sunlight are light above and darker below, where the shadow falls, and this pattern is used by predator and prey alike to detect animals at a distance. Reversing the pattern makes the animal less recognizable and the pattern of spots helps to disrupt its general outline. The fur, smooth and sleek, is typical of the cat family, and accentuates the streamlined body form.

Fish scales

Most fish are covered with scales (the most familiar exception being the eels which have thick, slimy, scaleless skins). Laid down in overlapping rows that spiral around the body of the fish, scales form a protective and streamlined frontier between the fish and its watery world. The size, shape, number, and arrangement of scales are among the main keys to the identification of a fish, since they vary hardly at all within most species – from the largest to the smallest individual, each has the same pattern of scales.

Fish scales come in a variety of styles. A few very primitive fish, including garpikes, have ganoid scales, diamond-shaped and forming a lattice pattern *(diagram a)*. They are covered with a hard, shiny material called ganoin. Such scales were characteristic of many prehistoric fishes, loading them with cumbersome, inflexible armour plating.

The typical bony fishes – trout, herring, cod, and hundreds of other species – have scales made of very thin, flake-like pieces of bone, often fine enough to be transparent. They are usually more or less rounded in outline: cycloid scales have smooth edges, while ctenoid scales have a spiked or serrated trailing edge *(diagram b)*. The scales grow in the dermis, the inner layer of the skin, and are covered by a fine epidermis or outer skin layer; each scale fits into its own little pocket of epidermis *(diagram c)*. The skin contains glands emitting mucus which keeps the scales slippery and flexible (as an angler knows to his cost) and also acts as an anti-septic, protecting the fish from bacterial infection. The scales grow by adding rings around the edge; they grow fast in summer but little in winter, and thus leave seasonal growth lines by which the age of the fish can be estimated. Scales taken at intervals from

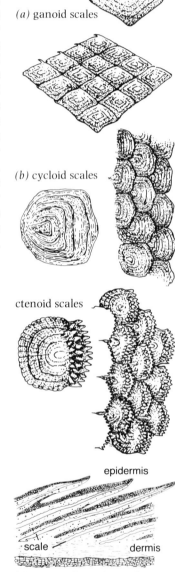

(a) ganoid scales

(b) cycloid scales

ctenoid scales

epidermis

scale   dermis

(c) Section through the skin of a bony fish.

A single cycloid scale from the skin of a bream, a freshwater fish with dull, inconspicuous colouring but reflective scales which mirror the light when viewed from certain angles. The ridges represent changes in growth-rate due to summer and winter conditions. This picture was taken using polaroid light and interference filters, so the colours are not true to life.

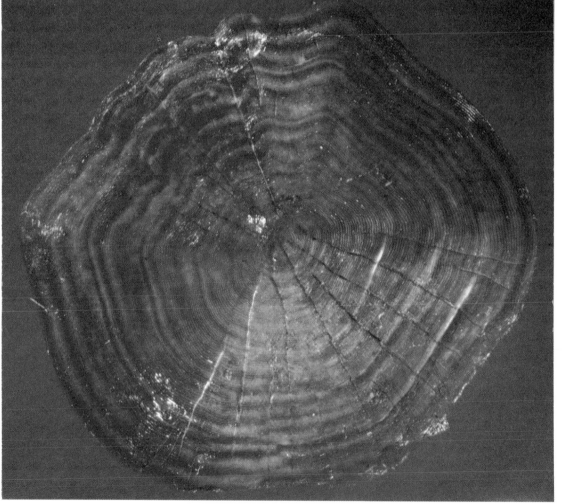

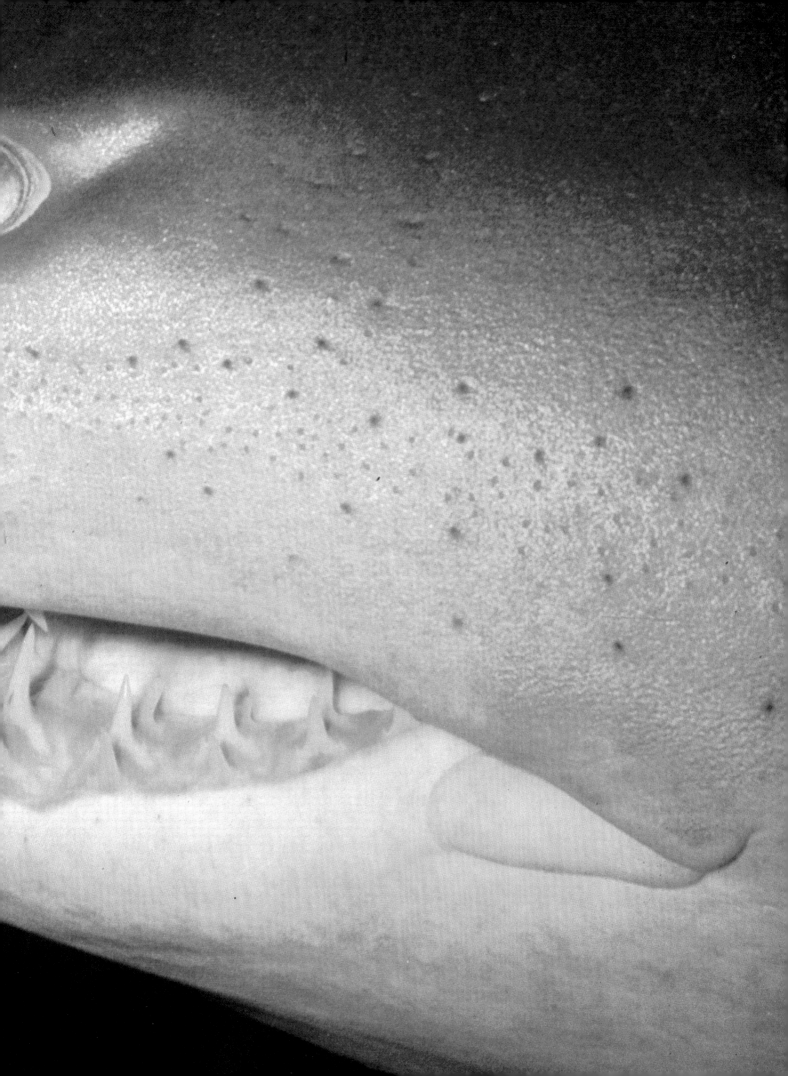

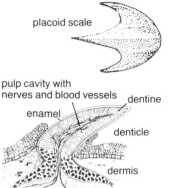

placoid scale

pulp cavity with
nerves and blood vessels

enamel

dentine

denticle

dermis

(d) Diagram and section through
a placoid scale.

A clownfish rests secure in a bed
of sea anemone tentacles. These
stinging tentacles are normally
used by the anemone to capture
fish prey, yet the clownfish
survives without damage and
actually thrives better in
association with the anemone
than on its own. Indeed,
clownfish have been observed
rubbing against the tentacles to
encourage the anemone to open,
and some scientists have
suggested that perhaps the fish
obtains a protective mucous
coating by this procedure.
Alternatively, it may be just
ridding itself of parasites. The
clownfish certainly gains
protection from predators by
living amongst this waving
weaponry, but it seems that the
anemone also gains: clownfish
have been seen driving away
fish that would prey on the
anemone, and some have even
been seen bringing food to the
anemone. It may also be that the
clownfish acts as a decoy, luring
other fish to their doom.

This pair of fish (*Pseudojuloides cerasinus* – no common name) from Australia illustrate the great disparity of colour that can arise between male and female. The male's colour probably serves as an attractant during courtship, and may also provide a territorial warning to other males. The female remains quite inconspicuous, so as to avoid predation until she has produced the next generation.

tagged wild fish provide even more evidence about fish migrations, health, and development.

The other main type of fish scales are those known as placoid scales or, more commonly and appropriately, denticles: 'little teeth' (*diagram d*). They are found on the primitive cartilaginous fishes, sharks, skates, and rays (whose skeletons are made of cartilage, not bone). Each denticle grows up from the dermis until its curved tip breaks the skin surface – denticles are not covered with skin as bony scales are. Each denticle, like a human tooth, is made of dentine (tooth ivory) capped with enamel; each has a pulp cavity containing nerves and blood vessels. Denticles are usually small, but may be sharp. Brushing against the skin of a shark, can flay the skin of a swimming man like a particularly vicious sandpaper. Sharkskin was indeed used as an abrasive by furniture craftsmen in the past. It may even be rubbed smooth and polished to produce a leather known as shagreen.

The fearsome teeth of sharks are actually modified denticles. Where the skin grows round the jaws, the denticles grow longer and sharper to form real teeth (*See pages 88–9*). Rows of them lie behind each other, and as the teeth are worn down or broken, new ones move forward to take their place. It is believed that all vertebrate teeth developed from denticles, but in most animals they became rooted in bony sockets in the jaw, instead of in the skin. Denticles are modified in other ways, too. The dagger-like spine in the tail of a sting ray is a type of denticle, with a groove in it through which the poison flows. Many of the skates and rays, however, have dispensed with ordinary denticles in the skin, leaving a lighter and more flexible body for swimming. The teeth in the long flat nose of the sawfish, shaped like a double-bladed saw, are also denticles. The sawfish thrashes its nose about in shoals of fish in the hope of stunning or injuring a few, to be picked up later at its leisure.

Reptile and mammal scales

Some people think that snakes are cold and slimy to the touch. They are wrong. Perhaps their belief arises from the term 'cold-blooded' which is applied to reptiles and many other animal groups; or from the glistening sheen of a healthy snake. Cold-blooded, however, simply means that the animal cannot raise its temperature by increased metabolic rate, but is more or less dependent on environmental sources of temperature regulation. A snake may feel quite warm to touch. And, of course, it is not slimy; its scales feel dry and silky, though somewhat rough if stroked against the grain.

Snakes and lizards, crocodiles and alligators, turtles, tortoises and terrapins are the major groups of reptiles living today. Almost all reptiles have scales. They are not like the scales of fishes (which are, indeed, cold and slimy). Reptile scales are outgrowths of the outer layer of skin, or epidermis. They are tough and horny, made mostly of the protein keratin, like our own fingernails. The skin runs between and around the scales in folds, keeping the body flexible (except in the tortoises). Sometimes the scales have an overlapping growth pattern like the tiles on a roof, as in the snakes. In

the crocodilians the scales may be large and plate-like, set flat in the skin. They are known as scutes, as are the fused scales that constitute the outer shell of most tortoises, turtles and terrapins.

The reptile's dermis, or underneath layer of skin, contains nerves and blood vessels, but few if any glands opening to the outside (such as sweat glands). This is a clue to the major function of the scaly skin: it is virtually waterproof, not to stop

An eyelash viper (or Schlegel's viper) lies in wait on a branch, ready to ambush a passing bird, which it will catch in flight, using its prehensile tail to anchor itself to the branch. It is a good illustration of how a coating of scales allows flexible body movements. The scales of a snake are important in its locomotion. Most snakes have a row of flat scales on their bellies, just visible in this picture, and these are moved forward, one after the other, digging into the substrate as the snake pulls its body forward.

A close-up of the side of a python, showing how scales can be used to make the exquisite mosaic patterns that have made such skins so highly prized for use as handbags and shoes. The python is not so agile as the eyelash viper. For an effective ambush at close quarters, a predator needs to be inconspicuous. The python's patterning of dark blotches surrounded by lighter areas serves the same purpose as the pattern of a giraffe, breaking up the body form – a 'dismembering' effect. It is particularly effective in the dappled light and shade of the jungle.

Crocodiles and alligators have rather different scales from those of other reptiles. Called 'scutes', they are bony and quite massive, but are not fused together or joined to the underlying skeleton, so flexible fast movement is still possible. Each scute develops on its own, and is replaced by layers from below. The scutes are particularly massive on the back, perhaps because this is the area most exposed to the sun and most at risk of drying out. As these pictures of an alligator show, where the scutes are largest, the area of less waterproof skin between is smallest, so large scutes provide a good seal against water loss. Areas of small scutes occur on the sides and around the shoulders and hips, where greater flexibility is needed during movement.

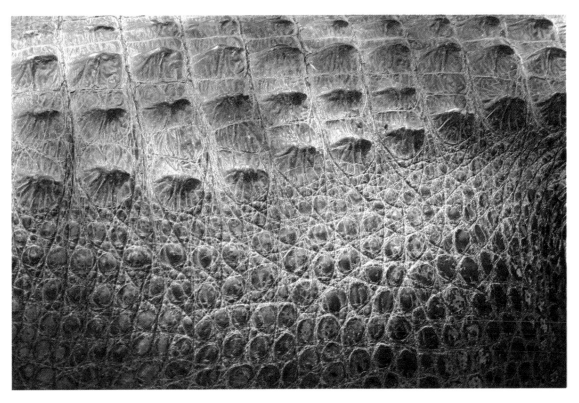

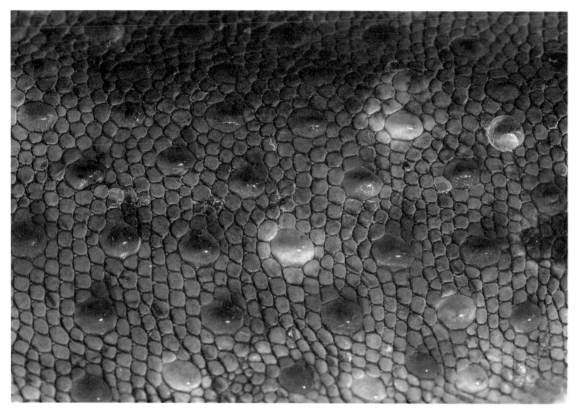

A sprinkling of knuckle-dusters in the form of colourful horny tubercles gives the Tokay gecko a rather formidable appearance. But the thin scales covering most of its body do not afford a great deal of protection against its enemies, and it is at its most active under cover of darkness.

water getting in but to prevent it escaping from the body. When the first reptiles evolved they were able to colonize the land, even its driest regions, by virtue of their scaly skins. Both reptiles and amphibians left their fishy forebears behind in the primeval waters; but most amphibians are still obliged to return to water to breed, and they must keep their thin skins moist with mucus. Reptiles, in their turn adapted scales to a land-based existence.

In many reptiles – excluding snakes – the scutes may be reinforced by bony plates, or osteoderms, within the dermis (*see diagram a*). These are taken a stage further by the crocodiles, which have large bony plates beneath their scutes; and to an extreme by the tortoises and turtles, whose spines are joined to bony plates that are fused together to form the familiar box-like shell around the animal. Over these bony plates lie the scutes, also fused together, though not at the same junctions as the osteoderms. The scutes of a tortoise grow by adding new layers at their edges, forming 'growth rings' (rather like the rings in a tree trunk) which can help to indicate the age of the animal. (They can certainly grow very old – a century or more in the case of the giant tortoises.) The scutes, tough but not brittle, can be easily and attractively worked by

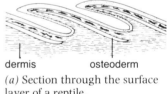

dermis   osteoderm

(*a*) Section through the surface layer of a reptile.

Quite a different covering and not at all flexible, the shell of a tortoise is made up of fused bony plates which are actually attached to the backbone, covered by fused dermal plates. These are the part of the shell that you actually see, and in this tortoise there is a high boss in the centre of each plate, marked out by a star-like pattern. The ridges on these plates represent annual growth fluctuations.

Not a piece of pine cone, but part of a pangolin! The pangolin or scaly anteater is the only mammal covered in scales. Made from modified hair (like the horn of a rhinoceros), the scales overlap like tiles from front to back, and are so arranged that the pangolin can roll itself up into an armour-plated ball when danger threatens, protecting its softer hairy underparts. Once it is rolled up, only a strong or skilful hunter, such as a lion or a man, can unroll it.

The rattle of a tiger rattlesnake, made up of a series of loosely connected horny segments. When the tail is moved, these make a noise like that of a hand-rattle; the rattlesnake uses it as a warning device to avoid unpleasant encounters with potential enemies or with larger animals who might accidentally tread on it. As the rattlesnake grows, it sheds its old tight skin from time to time, except at the tip of the tail, where the horny segments add a 'button' at each moult.

Butterfly scales

(Opposite) Part of a butterfly's wing magnified about 600 times to show the scales. Each scale is a modified hair, the flattened outgrowth of a single cell. Some scales are reduced to long hair-like structures. The scales overlap like roof tiles to completely cover the membranous surface of the wings. In the centre of the picture some scales have been removed to reveal the pockets in which they lie. Scales are not necessary for the wing to function – their main purpose is to give colour to the insect, but they are also said to help maintain a wing temperature suitable for efficient flight and to contribute to lift during flight. The colour of these scales is due not so much to pigment as to minute structural patterns on the scales which cause interference, diffraction or scattering of light. The two-tone colour of some of these scales is due to light being reflected from different angles of the scale structure.

Butterflies and moths are known to science as the Lepidoptera, a Greek word which means, literally, 'scaly wings'. The shimmering colours and vivid patterns of their wings are all created with scales. The scales are so small that they can only just be distinguished with the naked eye. They may be more easily noticed when a butterfly flaps its wings against an object, or flutters in the hand, leaving a powdering of what looks like fine dust, but is really scales. It is only under a microscope that one can appreciate their delicate beauty.

Each individual butterfly scale consists of a single dead cell, flattened and containing a tiny pocket of air. The scales are made of sclerotin, derived from the insect's cuticle. They are arranged in over-lapping rows like tiles on a roof. A typical scale is roughly rectangular in shape with rounded cor-ners. At the base of the scale is a short stem or peg which fits into a pocket on the wing membrane. The other end of the scale may be straight, rounded, wavy, broadly or finely serrated.

When a butterfly first emerges from its meta-morphosis in the pupa, its wings are soft bags covered with a layer of scales lying alongside each other like hundreds of packs of cards stood on end. As the wing expands and flattens out, the 'cards' – or scales – fan out and slide apart until they are lying in regular, overlapping rows. Meanwhile the clear wing cuticle dries out; it contains a network of hollow tubes or 'veins' which stiffen, supporting the wings rigidly like the struts in the fabric-covered wing of an early airplane.

The scales all point away from the leading edge of the wing, to help the air flow smoothly over the wings when the insect is in flight. (It has been calculated that scales provide 15 per cent more 'lift' to the butterfly and also improve its gliding performance.) If the scales are damaged or removed in any large quantity the resulting imbalance and uneven airflow will affect the insect's flight – although it will soon revise its flying technique to compensate for the damage.

Some butterfly scales are modified into short spiky points, others into long, fine hairs. In some butterflies, especially those known as the clear-wings, parts of the wing carry few or no scales, or scales reduced to tiny bristles. The effect of this is to camouflage the insect by allowing its background –

The heliconid butterfly *(left)* and the cabbage white *(right)* have just emerged from the chrysalis, and are slowly expanding their wings. Each has a colour pattern on its scales distinctive for its species, for its sex and perhaps even for its locality.

such as the flower on which it is feeding – to show through the wings. Clearwings are mainly mimics of bees, wasps and other noxious insects. Other species have thick hair-like scales on their wings and bodies: this is most noticeable among moths which fly by night and need the extra insulation of a fur coat. Butterflies are very sensitive to heat and control their temperature largely through their wings. By turning the wing surface at right angles to the sun's rays, the air in the scales and tracheae is warmed up and the heat transferred to the body. The colour of the scales also affects temperature: dark scales absorb sunlight, while shiny scales bounce it off.

The colours of the scales are produced in two ways: either by pigment cells produced by the insect or absorbed from its food plants; or by the structure of the scales, hence the name 'structural colours'. In the latter case, the surface of the scale is covered with a regular pattern of ridges and valleys, which can only be seen in detail under the scanning

This Venezuelan noctuid moth has a remarkable complex pattern on its wings. This helps to break up its outline and make it less noticeable to predators. The long fine pencil hairs on the hind wing are specialized for releasing pheromones (sex hormones) to attract a mate.

The iridescent pattern on the wing of a Madagascan Croesus moth is due to the fine structure of the scales causing interference patterns of reflected light; certain colours of light are reflected off microscopic ridges on the scales.

(Opposte) Part of the wing of a rare male purple emperor butterfly. Viewed from this angle it looks a glorious blue; from other angles it appears a duller brown. This can be confusing to a predator, since the butterfly appears to change colour as it flits in and out of the woodland sunlight: at one moment there is a magnificent blue butterfly, the next a rather dull inconspicuous one. The colours are a mixture of pigment and structural colours.

Like a watercolour painting, this bamboo page has softly patterned wings even when seen from the underside. When viewed from a distance, it is difficult to see exactly what shape this butterfly is. All these patterns are under strict genetic control: expression of the genes (factors which control heredity) carried by a particular species causes the localized synthesis of pigment molecules and/or of scales of a particular structural design. It is thought that the diffusion pattern of certain chemicals causing pigment synthesis are ultimately responsible for the pattern, which may help to explain why butterfly wings often resemble a colour wash.

A close-up of the 'eye' on the wing of an owl butterfly. An interesting effect of the pale ring around the central eye and the white markings on the black centre is to give the impression of light bouncing off a three-dimensional convex surface, like that of a real eye. As well as diverting an enemy's attack away from its own vulnerable head, the butterfly, by having such large 'eyes' on its wings, may make a predator mistake it for a much larger animal, and thus avoid attack altogether.

electron microscope. These ridges create colours by scattering light, to produce the metallic, iridescent, pearly or white effects on a butterfly's wings. Pigmented and structurally-coloured scales may be arranged in separate blocks or mixed together, producing a wonderful kaleidoscope of different optical effects.

The glorious colours and patterns of butterfly wings are very important in the lives of the insects. Patterns may be used for camouflage, warning, mimicry, deception, or mating signals. Apart from being able to distinguish all the colours of the spectrum that we can see, butterflies can also discern ultra-violet light which is invisible to us. Sometimes one species, for example, may mimic another so perfectly that we cannot immediately tell them apart; but another butterfly can do so by scale structures on the wings which reflect ultra-violet light patterns.

(Overleaf) A gulf fritillary flying past a cactus stem reveals the brilliant metallic silver patterning of its underwings. When viewed from this angle, its hind wings are much more brightly patterned than its forewings, yet when its wings are closed, both are seen to be equally highly patterned.

There is another curious function of Lepidoptera scales. In some species the males have scent-bearing scales, called androconia, on their wings. These may be scattered, or grouped as dark, hairy patches. When courting, the male dabs special 'brushes' at the end of his body on to these scent scales, and wafts the stimulating odour near the female to attract her. Some butterflies, such as the ithomids, cannot produce this scent until they have fed on certain plants. Exactly how these androconia work, and the chemicals involved, is still insufficiently understood.

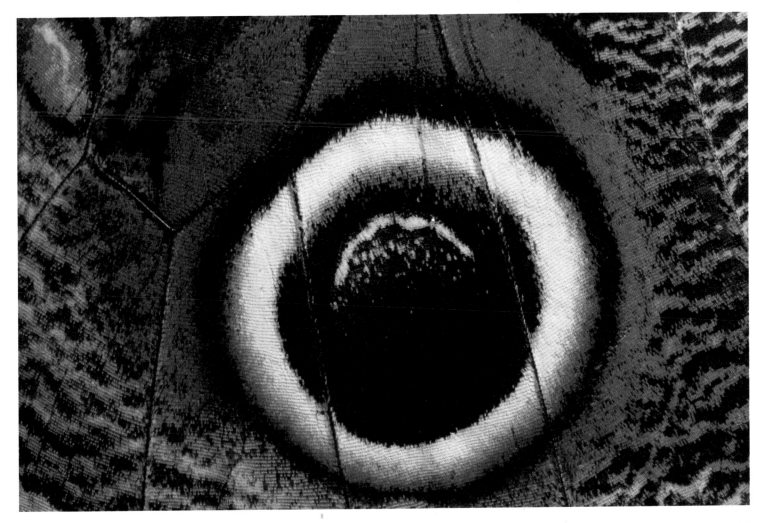

Feathers

Feathers are unique: they distinguish birds from all other animal groups. A hundred and fifty million years ago an avian ancestor, *Archaeopteryx*, a crow-sized flying creature, intermediate between reptiles and birds, had wing feathers that closely resembled those of modern birds. The origin of feathers is not known, but they probably evolved from reptile-type scales.

Birds are warm-blooded creatures and the primary use of their feathers is to allow them to keep their bodies at a constant temperature of 40–43°C, somewhat higher than ours; a bird will feel warm in one's hand. Feathers are excellent insulators, trapping layers of air next to the body. They also streamline the bird's body; some of the wing feathers are specially shaped and angled for this purpose (see page 39). Feathers also provide a

waterproof surface. They are made even more waterproof by being coated, during preening, with a thin film of oil from a gland near the bird's tail. Feathers are used in display and camouflage, with a tremendous versatility of design and colour – from the opulence of a cock pheasant to the neutral camouflage of a hen pheasant or winter white of a ptarmigan.

There are three basic types of feathers *(see diagram a)*: the outer contour feathers or pennae, including the flight feathers of wings and tail; the downy feathers, or plumulae, which lie beneath; and the filoplumes, fine hair-like feathers distributed among the others. They are all formed of keratin and grow from small pockets or follicles (the 'goose-bumps' apparent on a plucked fowl) in the skin of the bird.

The magnificent colouring of the head of a cock pheasant is related to his role as a mate. Male pheasants, once they have mated, take no further interest in either their mates or their offspring. Having no need to conceal himself while incubating the eggs, he can afford to evolve striking display colours.

filoplume

contour feather

contour feather

downy feather

(a) The three basic types of feather.

Feathers come in different shapes and sizes to suit their functions. *(Top)* The contour feathers of a pheasant, closely packed to give a smooth streamlined body covering. *(Right)* A body feather from a tawny owl, soft and fringed to muffle the sound of the owl's approach as it hunts. *(Far right)* The precisely structured 'eyed' feather from a peacock's tail, designed for display. Its shimmering colours are due to refraction and reflection of light from layers of horn which mask the underlying true colour, which is brown.

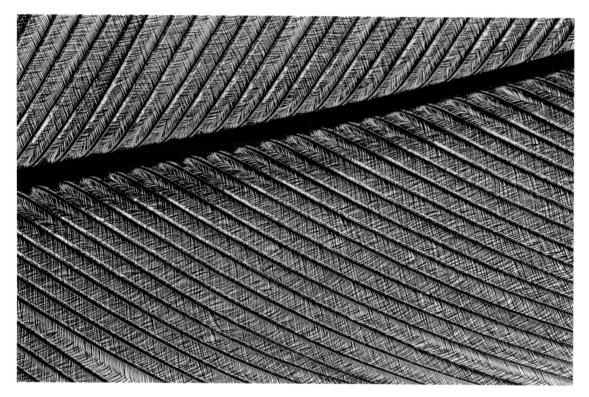

A close-up of a feather, showing the interlocked barbules forming a close meshwork between each barb.

A duckling dries itself in the sun after a swim, its feathers slightly bedraggled. At this age its feathers are all the downy type, needed to keep its tiny body warm.

One of the warmest nests in the world, the eider duck's nest is lined with soft down from the female eider duck's breast. Eider ducks nest on tundra-covered islands in the arctic, so it is not surprising that they have developed the warmest known down. When they hatch, the chicks already have their own eiderdown coats.

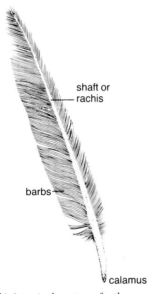

(b) A typical contour feather.

(c) Part of a bird's feather showing the main rachis, and the barbules linking the parallel barbs together.

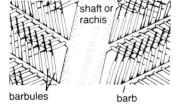

A king eider drake preening. His resplendent breeding plumage is lost shortly after breeding when he reverts to the more sober colours of the female.

Feathers have an elegant simplicity of design. A typical contour feather *(diagram b)* consists of a hollow base called the calamus and the shaft, or rachis, which gradually narrows and becomes more solid towards its tip. Along the shaft on either side arise hundreds of fine stems, or barbs. These, in turn, each bear many smaller filaments, or barbules. On one side of each barb, the barbules bear tiny hooks, while those on the other side do not; so, when the barbs lie in position parallel to each other *(diagram c)* the barbules interlock in a tight mesh or web. Most birds spend a fair amount of their time preening – drawing their feathers through their beaks to clean and tidy them. Simply by doing this they knit the barbules together, an action similar to that of closing a zip fastener.

The down feathers are small, with a short shaft and no hooked barbules, so the barbs do not interlock but fluff out in different directions. They can also be packed into a small space and re-expand afterwards to their original fluffiness. Down is the covering of small chicks and other nestlings, and provides excellent insulation, as anyone with a down-filled quilt can testify. The down of the eider duck has become synonymous with the best type of quilt. When the female eider duck nests in the cold springtime of its home in Scandinavia, she pulls tufts of thick down from beneath the contour feathers on her breast, and uses the down to make a thick, warm lining for her nest. The down is taken

The vulture looks like some avian dignitary with a ruffed cloak. The feathers on its head and neck are very short and downy, so that it can put its head right inside carcasses to feed without getting its plumage into too much of a gory mess. The long neck also helps it to reach into carcasses.

The penguins *(below)* lost the power of flight some 100 million years ago, and have no flight feathers on their wings. Their stiff close-packed feathers form a thick insulating mat that is impervious to water and provides a good streamlined surface for swimming. Penguins of the genus *Eudyptes (far right)* have a distinctive plume of yellowish feathers on either side of the head. Such minor adornment is nothing compared to the head feathers of the crowned crane *(right)*, present in both male and female, which serve no obvious purpose.

from the nests by local people and used to stuff quilts or 'eiderdowns'. Down is extremely light, making it ideal for cold-climate clothing and equipment; down-filled suits and sleeping bags have revolutionized mountaineering and polar exploration. Despite the proverbial lightness of feathers, however, the plumage of a bird may weigh twice as much as its bony skeleton – as much as 14 per cent of the total weight of an eagle, for example.

The contour feathers start to grow when a chick is a few weeks old. The larger the bird, the larger and more numerous its feathers: the crow may have 2000–3000, while a swan may have as many as 25 000 in winter – when plumage is always thicker. The feathers are moulted and replaced at least once a year. Many birds are forced to remain flightless, and hence more vulnerable, while they moult their flight feathers.

Some birds do not fly at all, and bear fewer, or modified, flight feathers. Penguins, for instance, have taken to swimming instead of flying, and have a thick coat of mostly downy feathers in which they can withstand even the coldest Antarctic winters. By contrast, the ostrich in the heat of Africa does not need feathers for warmth, but to keep cool: its feathers consist largely of long, soft, unhooked plumes. Like those of the egret, these plumes were greatly in demand in bygone days for decorating women's millinery and clothing.

Shells

This thorny oyster from the Red Sea reef may be as much as 20 cm in diameter, and is seen here in its favourite position, attached to a rocky reef overhang. Its bright red colouring is due to a covering of a red sponge of the genus *Microciona*, which is often found on the upper valve of this oyster.

Molluscs are a group of animals that includes snails and slugs, octopuses, squid and cuttlefish, and the nautilus. Of these, cuttlefish hide their shells inside their bodies and octopuses and most squid have lost them, as have the slugs. The shelled molluscs, however, still form one of the most numerous, diverse and successful groups of animal species. They are one of the oldest groups, dating back over 600 million years of our planet's history. They have also literally shaped our world, for many rocks not only contain fossil mollusc shells but are actually formed of them. Shells are amazingly durable; we would know little of the ammonites and others that roamed the primeval oceans if they had not borne shells.

Mollusc shells have also played a part in shaping our cultures. The pearls produced by the pearl oyster are among the most valued of jewels. And in the islands and countries around the Indian Ocean, cowrie shells were used as money. Explorers and colonizers were happy to adopt this form of currency, to the extent that in the nineteenth century it was worth the while of the expanding British Empire to bring cowries from India to England and then send them out to Africa, to placate the natives of the violated continent.

These shells which have had such an influence on our lives range in size from millimetres to nearly 1.5 m in length, in the case of the giant clam shell. Although some mollusc shells are similar in form

A cone shell's attractive yellow and black coloration should be taken as a warning: this is one of the most poisonous shellfish in the world. A bite from the hollow teeth of its radula can inject enough venom to kill a man. The tooth is detached and shot into the prey like a harpoon, armed with a powerful nerve poison.

In the warm waters of Australia's Great Barrier Reef, a giant clam feeds, its valves agape to filter water and extract the tiny living organisms from it. It may eventually reach a length of 1.8 metres and can weigh up to 200 kg. The massive shell affords a home for encrusting sponges and corals, but even more remarkable is its living farm of algae, tiny plants which trap sunlight and use its energy to make living tissues. The striking blue colour of the soft tissues lining its fluted shell is due to millions of tiny algae living in clusters inside special cells. These are 'harvested' by being digested by special cells in the clam's blood. The clam in effect takes care of its crop by controlling the light intensity to which the algae are exposed – it uses special lenses to focus light on the algae or, if the light is too bright, it extends a pigment screen to protect them.

Two Australian cowries (*left*) move slowly over the reef. Their heads and mantle are extruded through the narrow slit on the underside of the shell. Another Australian cowrie (*below*) shows the extent of the mantle, which in some species is curled back over the shell to make it too slippery for a predator to get hold of. From time immemorial cowries have been used in many parts of the world to make necklaces and ornaments, and are also believed to have magical properites. Because of their varied and attractive shells, they are also collector's items, and can fetch very high prices. From early times, cowries were used as currency in parts of Africa, and used to be imported to England from the Indian Ocean and re-exported to Africa by the tonne.

A hermit crab stands hopefully by the sea anemone of its choice, trying to perusade the anemone to move on to its shell. This anemone is sometimes called a 'parasitic anemone' because it is so often found riding on a hermit crab. The association seems to be of mutual benefit: the anemone feeds from scraps left over from the crab's meals, and is transported to new feeding grounds by the crab, which gains extra protection from its enemies by the anemone's stinging tentacles.

A hermit crab emerges from its shell to feed. These crabs have soft bodies and live in 'borrowed' mollusc shells, moving to a larger shell from time to time as they outgrow the existing one. When danger threatens, the crab can withdraw deep into the shell, blocking the hole with its larger claw. In the coastal waters in which hermit crabs live, the borrowed shell is thought to give better protection against the buffeting of the waves (as well as against predators) than would a normal carapace. This perhaps compensates for having to drag a heavier shell around.

Eyes

The face of an animal is rather like a control panel where several vital, specialized and sensitive instruments are gathered together. The human face comprises most of these instruments: eyes, ears, nose and mouth with teeth and tongue. More finely tuned equivalents of each may be found in the animal kingdom, as well as other instruments of which we have none, such as antennae. The capacities of our control panel may be limited, but if any one of these senses fails for any reason we feel its loss acutely. Functioning together, almost incessantly, they provide us with an extremely detailed and accurate assessment of the world and our place in it at any given moment.

In many animals, including ourselves, the dominant facial feature is the eyes. Their task is to translate light into information about the animal's surroundings. In the simplest terms this is done by admitting light into the eye (usually through a transparent lens) to fall on light-sensitive pigments contained in a group of cells called a retina. This causes a chemical reaction that electrically stimulates nerve endings, which in turn pass a message to the visual centre in the brain where it is translated into some kind of image. Whether that image is just an awareness of light, or a flickering jumble of colours, or a precisely detailed picture, is all in the eye of the beholder.

Human beings are so dependent on eyes that they can hardly imagine living in an invisible world. Some animals, however, have little or no need of eyes. Moles and earthworms tunnel blindly through the soil; eyeless cave fishes swim happily upside-down under rock ledges; a bat navigates in pitch darkness by bouncing sound waves off objects and listening for the echoes. On the other hand, many animals have eyes that are capable of forming very detailed images of very distant objects, even more acutely than we can. We could never emulate the hawk which, from 30 m up in the air, can distinguish a tiny mouse amid the shimmering grass, nor the gannet which can dive into the ocean to catch a fish whose precise position is disguised by the refraction of light under water.

Although vision is not indispensable to a full and happy life, most animals have eyes of some kind, whether rudimentary or sophisticated. Even tiny planktonic creatures may possess one or two eyespots by which they are able to discern light. As it is believed that the vertical migrations of sea plankton may occur in response to the amount of light penetrating the water, even this basic level of vision is important. Other simple creatures such as flatworms have eye-spots, a few retinal cells with no lens to concentrate the light or make an image. Many insects and other invertebrates, and even

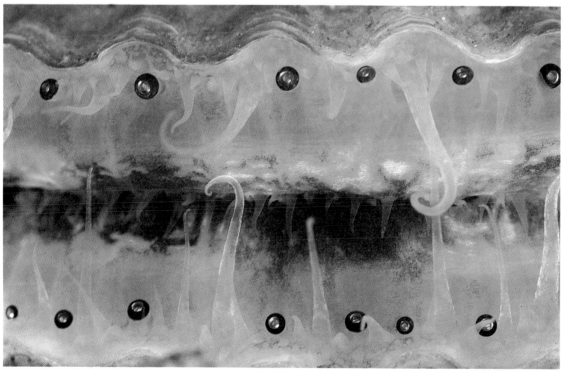

The eyes of a queen scallop are dotted all around the edge of its mantle. The jewel-like effect is due to a reflecting layer or tapetum behind each eye. Scallop eyes contain two types of retina – one responds to light, the other to sudden darkness, such as the shadow cast by an approaching predator. The scallop probably cannot interpret shapes, but can detect changing patterns of movement, such as moving light-dark changes. The eyes are not its only source of warning – the tentacles around the mantle edge are extremely sensitive to certain chemicals, and can probably detect the approach of a starfish long before its shadow falls, certainly in time for the scallop to close its valves or leap away.

(*Overleaf*) Three red-eyed tree frogs watch eagerly as their picture is taken. Their eyes are large and bulging, affording a wide angle of view and a certain amount of binocular vision for ambushing passing insect prey.

Eyes for all occasions. The jumping spider *(top)* has simple eyes almost all around its head to detect the approach of prey, with particularly large ones at the front to help it judge distance when pouncing on its prey. The tree frog eye *(lower left)*, used particularly in dim light at dusk or even at night, has a pupil which can be closed down to a slit by day to protect the visual cells from strong light. This is a vertebrate eye, the most highly developed kind found in the animal kingdom, giving the most acute vision. The horsefly *(lower right)* has the compound eyes typical of insects. These are a mosaic of tiny units called ommatidia, each producing a separate image, so that the overall picture received by the insect is a two-dimensional dot picture, rather like a magnified newspaper dot picture.

higher animals such as frogs, are sensitive to light all over their bodies, although they cannot 'see' with their skin. The scallop is the record-holder for sheer numbers of eyes. It may have from 50 to 200 simple eyes, strung along the edge of its mantle like a string of glistening beads.

Among the arthropods (including crabs, insects and spiders) there are two main types of eye: 'simple eyes' or ocelli, and compound eyes. Many insects

and some other arthropods have more than one type of eye on the face, each kind fulfilling a different function. The total number of eyes is another variable – two compound eyes and three ocelli is a common pattern among insects, but extra ocelli may be placed elsewhere on the body; and most spiders have eight separate ocelli of different sizes on the front of the head, but no compound eyes.

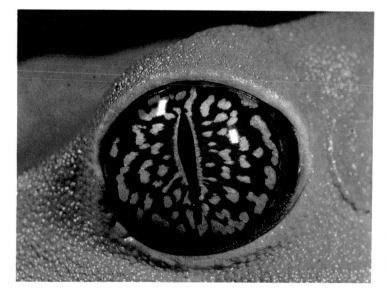

Each ocellus usually consists of a small lens backed up by several pigmented retinal cells, which can determine the quality and source of light and usually perceive something moving nearby. Ocelli usually look like small dark dots, and are often grouped in a triangle on the back of an insect's head. They enable the insect to judge the length of daylight, for example, by which it may regulate its whole life cycle. Spiders' eyes form extremely good images and have, for their size, excellent resolution.

Compound eyes are more complex, and usually more efficient – though still only a fraction as efficient as good vertebrate eyes. Compound eyes are the two large rounded or oval bumps on the heads of many insects, sometimes so large and prominent that they cover most of the head. A compound eye is really a collection of tiny visual units, known as ommatidia, each with a very limited visual capacity. The ommatidia are long, narrow and tube-like, slightly conical in shape. They are bunched together tightly like a bundle of drinking straws. One compound eye may comprise anything from less than ten to more than 25 000 ommatidia, depending on whose face it is attached to: a housefly, for example, has about 4000, while a dragonfly may have up to 28 000 ommatidia per compound eye. Generally, the more ommatidia the

better the vision, so it is not surprising that dragonflies have the most acute vision among arthropods.

Each ommatidium (*diagram a*) consists of several basic parts. There is a layer of transparent cuticle on the outside, which allows light into a lens beneath it. This is usually surrounded by cells containing 'scattering pigment' which absorbs scattered or incidental light rays, so that the only light entering the ommatidium is directly parallel to its axis. This beam of light is directed by the lens down the narrow visual centre or rhabdom where it reacts with pigment, stimulating the nerve cells that surround the rhabdom. The nerve cells pass the message to the optical centre in the insect's 'brain' where it is interpreted.

The ommatidia of different insects are varied. They may even be of different sizes within a single compound eye. The scattering pigment reduces the total amount of light entering the eye, so insects active by day may find themselves blind at dusk when the light is lower and more diffused. Nocturnal insects, however, often have the ability to withdraw the scattering pigment from their eyes at night in order to absorb every scrap of available light and to allow light from many of the lens facets to focus on a single light-sensitive rhabdom, thus

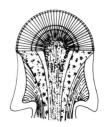

(*a*) Section through a typical compound eye with detail (*below*) of an individual ommatidium. The number of ommatidia in a compound eye varies for different insect groups – from as few as ten to more than 25 000.

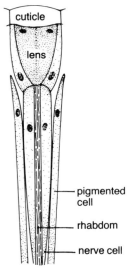

cuticle

lens

pigmented cell

rhabdom

nerve cell

The *Aeshna* dragonfly has very large compound eyes resembling a pair of headphones. It is a voracious hunter, catching other insects on the wing, so it needs very accurate vision. The eyes are quite far apart, which gives it good binocular vision for judging distance. As it approaches a potential prey animal, the image of the prey on the retinas of the two eyes moves towards the midline of the insect, and the dragonfly knows by experience just where on the retinas the image should be when the prey can be seized.

The eye of a diamondback moth seen under the electron microscope. The mosaic of hexagonal ommatidia can be seen clearly, magnified about 250 times. A ring of pigment cells separates each ommatidium from its neighbours, screening them from light received by adjacent ommatidia.

increasing the effective aperture of the lens system. Many moths go even further, possessing (like cats and some other animals) a kind of mirror – the tapetum – at the back of the eye: this reflects light back through the retinal cells, so every beam of light is used twice over.

Nobody knows what an insect really sees, but it is probably a multitude of pieces of light and colour of various shades and intensities – like a mosaic, or a kaleidoscope. It may also be compared to a poorly-printed colour picture in a book, where the dots of colour are separate and out of focus. Images, especially at a distance, must be very indistinct –

just a blur. The compound eye is, however, very sensitive to movement. The beams of light reflected from a moving object will fall on different ommatidia in succession, producing a series of visual stimuli. When one considers that, in some species, each compound eye can make an arc of more than 180° round one side of the insect's head – the two eyes therefore giving it all-round vision – it is not surprising how hard it is to creep up unnoticed behind a fly and swat it with a newspaper. A large dragonfly can distinguish movement as far as 10 m away, but even if its prey is much closer, the dragonfly may well lose sight of it if the prey sits

quite still. The whirligig water beetle, which often hunts on the surface of a pond, has compound eyes which are adapted like bifocal glasses to see both upwards into the air and downwards below the water surface. Another arthropod method of achieving all-round vision is to have eyes on mobile stalks, like the ghost crab.

A great many insects obtain their food from plants – especially flowers – so inevitably they have good colour vision; bees and butterflies can distinguish a wide variety of colours. Not only can they see most of the colours of the spectrum that we can, but they also see ultra-violet wavelengths. To us, a flower may look simply yellow with a black central dot, but to a bee it may have stripes, splashes and dots of ultra-violet pattern. (These ultra-violet markings on a flower usually radiate from the centre, directing the insect to the flower's sex organs to ensure pollination.) Ultra-violet patterns occur not only in flowers but also on the insects themselves. The wings of moths and butterflies may have ultra-violet markings that we cannot see. Male and female luna moths, for example, may look pale green to us, but in ultra-violet light the male is dark and the female pale. Some butterfly species mimic other species in coloration so closely that we cannot tell them apart, but they can distinguish each other by ultra-violet markings.

In some ways the eyes of vertebrates are easier to understand than those of invertebrates, for we can use our own eyes as a typical model. The eye resembles a camera. It consists (diagram b) of an almost completely spherical ball cupped in a socket in the skull, where it is often cushioned with fat and constantly bathed in fluid. The eyeball is formed of several tough layers kept taut by a filling of clear

jelly-like substance, the 'vitreous humour'. Light rays enter the front of the eye through a transparent window, the cornea, and a small hole, the pupil, whose size is regulated by the expansion or contraction of the circular iris. The light admitted through the pupil enters a clear lens, which focuses it on to the retina – the inner surface of the back of the eyeball. Here the focused image reacts with visual pigment, and the resulting nerve impulses are passed to the visual centre of the brain for decoding.

The clarity, detail and colour of the image depend on several factors, including the shape of the eyeball and the lens; the capacity of the eye to focus, by changing the shape or position of the lens; and the presence and number of two types of cells in the retina – the rods and the cones, so called because of their approximate shape. The rods react very sensitively to light, so they are particularly useful when there is little light available; but they cannot interpret colour. The cones require much more light to cause a reaction, but they are responsible for colour vision. Some animals, especially the nocturnal ones, have mostly rod cells in the retina; other animals, active by day, have mostly cones. Many animals, including human beings, have a mixture of both.

Being a particularly decliate instrument, the eye needs protection – usually, an eyelid. Most mammals have two eyelids, one above and one below, but some – such as horses and deer – have a third, inner eyelid, the nictitating membrane, which may move upwards or sideways across the eyeball. Both types of eyelid can be closed to protect the eye from a blow, or from dirt; in closing – blinking – they wipe the eyeball clean and lubricate it with teardrops.

Having eyes on stalks is particularly useful for the ghost crab, which spends much of its time buried in the sand ready to ambush passing animals. Only its eyes protrude like a pair of periscopes to scan the beach for the next meal. Well above its head and body, they command an excellent field of view.

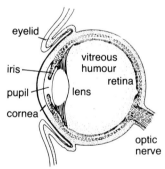

(b) Cross-section of a human eye.

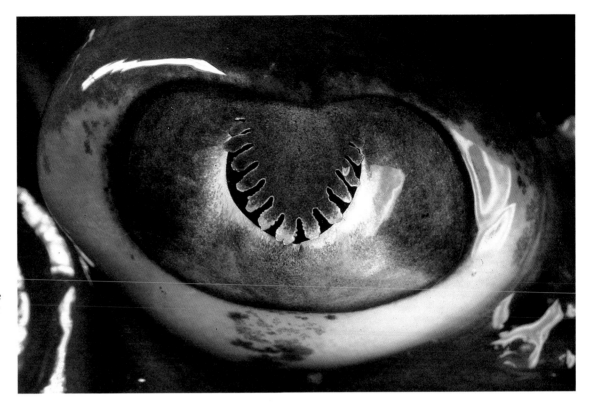

Close-up of the eye of a ray. The ray's eye is built along a similar plan to that of the human eye, in the form of a camera with sensitive retina, a focusing lens and an iris. The ray has very large eyes and uses them to locate its prey.

(Far right) A mudskipper lies in the warm shallow water of the Persian Gulf, keeping an eye open for food and foes. It can be almost totally submerged, yet still have its eyes above water.

(Below) Close-up of a horse's eye. The nictitating membrane, or third eyelid is partially visible in the right-hand corner of the eye.

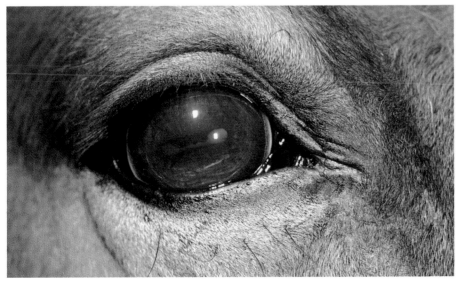

Some diving birds close the nictitating membrane when they are chasing prey underwater, but they can see through it – the membrane alters the focus of the eye to adapt to underwater light effects.

A great deal can be learned about an animal by the position of its eyes. Where the fields of vision of each eye cross in front of the animal *(diagram c)* the result is 'binocular vision' – the brain receives two slightly different images and superimposes them to produce a much more defined, three-dimensional impression. The further forward on the face the eyes are placed, the greater the degree of binocular vision. Animals with forward-directed eyes are usually predatory: they need accurate vision and judgement of distance in order to catch their prey.

The prey animals, on the other hand, need as wide a field of vision as possible so that they can see danger approaching from any side, even at the expense of accurate forward vision: so their eyes are set at the sides of the face, often bulging out. A hare, the popular prey of several creatures, can see behind it and judge how closely it is being chased (unfortunately hares have been known to run over cliff tops apparently because they were looking behind themselves instead of forwards). The difference in predator/prey eye positions can also be seen among fishes. Fish with large side-set eyes, like the carp, are preyed upon by fish with forward-placed eyes, like the pike.

One fish that does not fit either pattern is the mudskipper, whose eyes are set right on top of its head like twin periscopes. The mudskipper can lie in shallow water with its whole body submerged save for its eyes, which scan the surface and the water's edge for food. In fact its eyes are adapted for vision in air and it cannot even focus properly underwater – an odd situation for a fish. The so-called four-eyed fish has a similar problem but has solved it by having two pupils in each eye to control the light entering the lens: the upper part of the eye is designed to see in air, the lower ones to see in water.

The chameleon can look in more than one direction at once: its eyes move independently. The cornea is right at the tip of a protruding cone of skin, and the eyes can swivel about 180° in almost all directions. Only a fly perched on the top of a chameleons's head could escape these all-seeing eyes. The chameleon waits, motionless and camouflaged, until a prey insect comes within reach, then it shoots out its long sticky tongue to catch it. At the moment of attack, both eyes are directed forward to give the stereoscopic vision necessary for an accurate strike.

(Far right) The eyes of a hunter: the saw whet owl. Placed right at the front of the head for accurate stereoscopic vision, these eyes are not the owl's only means of locating prey. It relies also on its very sensitive hearing, and the rings of flattened feathers which accentuate its eyes are actually designed to funnel sound. The large pupils are an adaptation to a nocturnal life, allowing as much light as possible to reach the retina.

The Nile crocodile has eyes situated in bumps on top of its head, so that it can lurk unseen, with only its eyes and nostrils above the water, to wait for unsuspecting animals to come to the water's edge to drink. The ear is also visible in this picture, a horizontal slit behind the eye. The eye has a transparent membrane, the nictitating membrane, which can be moved over it when the crocodile is submerged, protecting it while allowing it to see underwater.

Several aquatic mammals also have eyes on top of the head. The hippopotamus, preferring to spend the hot day submerged in a muddy pool, has nostrils and eyes set on top of its head so that it can see and breathe while keeping its whole body underwater. The largest of rodents, the South American capybara, has a similar taste for wallowing in water and a similar positioning of eyes and nose. The crocodile is another example: it lies in wait beside a watering-place, and all that the prey – such as an unfortunate gazelle – can see is a couple of bump-like bubbles floating on the water, which are the crocodile's unblinking, menacing eyes. The gazelle does not notice the 'bubbles' floating slowly closer until there is a swirl, a snap, and a huge splash, as the gazelle is dragged to a watery grave.

Perhaps the strangest of animal eyes belong to the chameleon. They are mounted in twin conical turrets and can move independently of each other, giving the chameleon the ability to see all round itself when seeking prey, and binocular vision in front when it is preparing to strike with its long, sticky tongue.

The size of eyes is another clue to their owner's lifestyle. Very large eyes in proportion to the rest of the face almost invariably indicate a nocturnal animal. The largest eyes relative to head size belong to the tarsier, a small, shy, monkey-like primate which is active at night in the forests of Malaysia. Most owls, too, have very large eyes, set well forward for binocular hunting vision. By day the pupil is almost closed, but at night it opens wide to

admit maximum light. The owl's retinal cells are composed mainly of rods, giving the owl a night-time vision a hundred times keener than our own. Most birds have only cones in the retina and cannot see in the dark. The owl's nocturnal vision is used mainly to navigate through trees or buildings in darkness, for it hunts as much by its acute hearing *(see page 167)* as by its sight. Where the eyes of an animal are very small one can expect to find another sense organ taking the place of vision. The smallest eyes of any vertebrate are found in the mole. They are the size of a pinhead and buried in fur; useless as image-forming eyes when in the mole's subterranean tunnels but important as light detectors. The mole seeks its prey of worms and small animals with its extremely sensitive nose, but still needs eyes to detect when it has broken out of the burrow. Other types of animals which have taken to living like moles in underground burrows, have also lost their powers of vision: the marsupial mole and the mole rat, neither related to true moles, both have minute and comparatively useless eyes quite unlike the eyes of other marsupials and other rats.

Antennae

The antennae ('feelers') of an insect and the trunk of an elephant may be very different in size and appearance but they are equally versatile. Both are mainly organs of smell that have also evolved to perform many other tasks. The uses of an elephant's trunk are varied; the antennae of insects may be used perhaps even more diversely and certainly more sensitively, for smelling, touching, communication of messages, detecting wind speed, hearing, gauging temperature and humidity, tasting, even grasping things. Of all these functions, however, smell is usually the most important. In insects it can be incredibly acute.

Antennae come in a range of shapes and sizes *(diagram a)*. It is safe enough to generalize that the larger and more complex the antennae, the more important is the sense of smell to the insect. But a lack of visible antennae does not necessarily mean that the insect has no sense of smell; rather that its olfactory (smelling) cells are located elsewhere on its body. The antennae of a typical insect are covered with thousands of sensory cells called sensilla, adapted for smelling – receiving and interpreting airborne chemicals. The sensilla usually take the form of short, hollow, hair-like structures formed of the insect cuticle. There are

(a) A selection of insect antennae showing the variety of shapes and sizes.

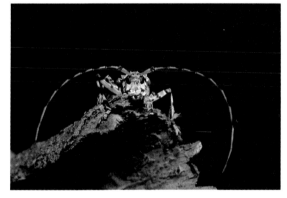

Antennae come in a wide variety of shapes and sizes – fine and smooth in the lacewing *(centre left)*, thick and inconspicuous in the spider-hunting wasp *(top left)*, or conspicuously jointed in long-horn beetles *(center and lower right)*. Large antennae, such as those of the cockchafer *(lower left)*, are usually very sensitive to smell. The male atlas moth *(top right)* can detect the odour of a female up to 2 km away.

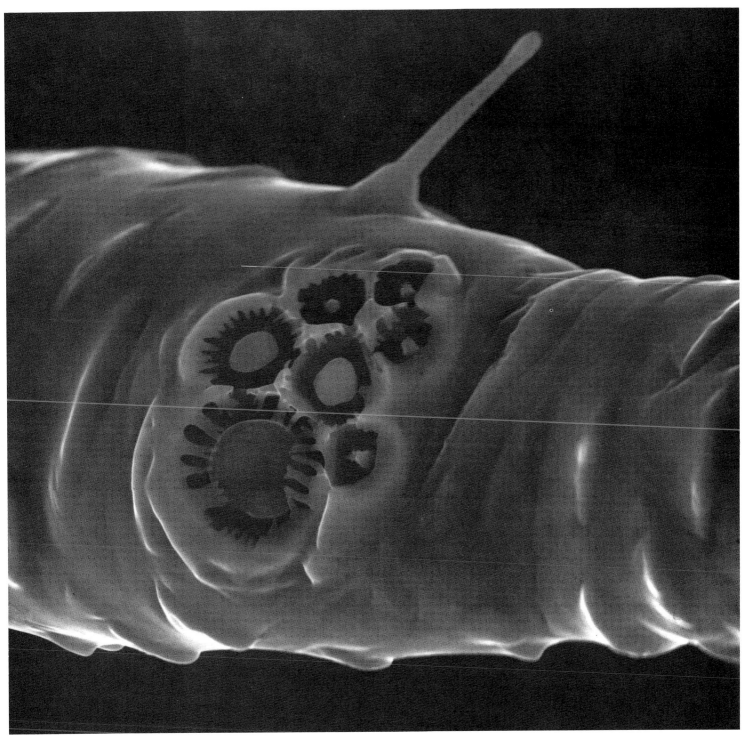

A cluster of sensors (called sensillae) on an aphid's antenna, magnified several thousand times by the electron microscope. Such sensors are scattered all over the bodies of insects, but are most numerous on the appendages. Sensillae are thought to be derived from hair cells and each comprises a receptor cell plus a group of accessory cells.

tiny openings or pores on the surface of the sensilla which admit odour molecules carried in the air. These trigger chemical receptor cells in the base of the sensilla which in turn pass a message to the insect's 'brain' where the smell is identified. The larger the surface area of the antennae, the greater the number of sensilla and the better the sense of smell. A honeybee may have 30 000 sensillae on each antenna, and a male moth has many millions of them.

The sense of smell is vital to most insects for finding food. Flies are particularly sensitive to the chemical odour given off by rotting meat, for example, while young bees must learn to recognize the scent of the best nectar-producing flowers. The scent of most flowers is deliberately produced to attract insects that will pollinate the flowers while they search for their reward of nectar.

Insects also use their antennae for smelling out suitable egg-laying sites, and these may well include plants that the adult insects never eat. The egg-laying female butterfly knows by instinct what kind of leaves the caterpillars will want to eat when they hatch. A female parasitic wood wasp cannot only smell a grub within wood, but can also detect whether another female wasp has already laid an

egg on the grub. A chafer beetle in mid-flight may stop and dive to earth as it smells the odour of a truffle, growing out of sight under the earth.

Social insects may lay scent trails. Ants can follow invisible but identical paths, their antennae picking up chemical markers left by the trail-blazer. As they scuttle back and forth along a food-supply trail they stop and communicate information to each other by tapping with their antennae. Each individual hive of honeybees, too, has its own characteristic smell, but guard bees at the hive entrance will not admit any bee that does not carry the hive's personal perfume.

(Far left) The female parasitic woodwasp lays her egg on the grub of another insect already inside the tree trunk. Her antennae can detect not only the presence of the grub but also whether another wasp has already laid her egg there.

The pollen-laden honeybee in the centre has just returned from a successful foraging trip. The surrounding workers are using their antennae to learn the smell of the flowers she has visited, so that they can also visit them.

Three wood ants communicating. By tapping with their antennae, ants can convey information to each other about food sources in the area, the direction of foraging trails, and so on. Their acute sense of smell can also discriminate between ants of the same species from different colonies.

This cockchafer, or maybug, has short antennae ending in movable clubs which it can spread out like a fan, greatly increasing the area of sensillae exposed to the surrounding air. It feeds on nectar and other plant food, and can even detect the presence of an underground mushroom as it flies over it. A heavy insect, it uses internal air sacs to help it stay airborne.

Most insects have a 'brain' that is not nearly large or complex enough to be capable of anything resembling reason. They are like mini-robots, genetically programmed to perform one or two basic tasks, after which their brief lives are expendable. Often odours set these life cycles in motion. Such smells are called pheromones, produced by a species as a signal to its own kind to do something specific. The most potent of these pheromones are concerned with sex. Insects are so small that if they left sex to chance encounters they would soon die out. So in some cases, males emit sex pheromones that attract females; in other cases the females' scent attracts the males.

Perhaps the most remarkable example of this scent seduction occurs in the silk moth *Bombyx mori*. The female moth sheds a tiny amount of scent on to the wind (in a laboratory experiment it took a third of a million female moths to produce one small

A male cherry moth *(left)* and a male tussor silk moth *(below)* show the large and much branched antennae characteristic of male moths. This structure gives a very large surface area of antennae over which to distribute sensillae – tiny sensors which can detect minute quantities of chemicals wafting on the air. The male moth needs these giant aerial-like antennae to find a mate. When ready to mate, the female moth emits a scented sex hormone (pheromone). Borne on the air currents, it penetrates every crevice where a male might be, and stimulates him to search for her, flying down the concentration of perfume to find her.

The head of a newly emerged male mosquito, magnified 85 times, showing the long central proboscis flanked by two maxillary palps, then the brush-like antennae. These are also used to locate a mate, but this time they are to detect sound waves. The sensillae on these antennae have long fine bristles which pick up the vibrations in the air set up by the female's wingbeats. This arouses his sexual instincts, and he flies towards the sound to find a mate. The antennae are very selective – the male mosquito responds only to the vibration frequency of the female wingbeats, and will not fly towards other males.

drop of the pure pheromone). The male silk moth has a pair of the most dramatic antennae in the insect world, elaborate plumes that sense the wind as a radio telescope bowl scans the vast reaches of space. The male moth has only one aim and purpose in life: to mate with a female. He has no mouthparts and cannot eat, and he must fulfil his destiny in a short lifespan of a week or two. It has been calculated that a male silk moth can trace a female more than 2 km away, and that just one single infinitesimal molecule of her scent landing on one of his antennal sensilla is enough to set him on his vital journey. Since the female's scent is blown downwind, the male finds her by flying upwind. He follows the increasing concentration of her scent, and directs his flight according to how much of it reacts with each antenna: a moth with only one antenna can detect the female's scent, but cannot accurately follow it to its source.

The antennae of male mosquitoes and midges are also adapted to find females of their kind, but in a different way. The brush-like male antennae are sensitive to sound waves, particularly to those of the frequency of the wingbeat of females; so when a male midge 'hears' a female with his antennae, he flies towards the source of the good vibrations. The response is so simple that the insect will be attracted to anything producing vibrations at the correct frequency (even a tuning-fork), and scientists are investigating ways of using this instinct as a type of insecticide by distracting the males from mating with the females.

Noses

The importance of smell in the life of many mammals is illustrated by the behaviour of pet dogs. This Alsatian (German Shepherd) dog is sniffing the ground to find out who has been here before him. A dog's nose is probably as important as its sight as a source of information. Watch a dog out for a walk: it will sniff the droppings of any dog that has passed by, to find out who else is in its territory. If it meets another dog, it will sniff around its tail, where the scent is strongest, so that if it passes that dog's trail again it will be familiar.

Most animals literally follow their noses. The face, equipped with its various sense organs to monitor the environment ahead, precedes the body; and very often the nose precedes the rest of the face. Such is the importance of a nose – or, rather, of the primary task of the nose, which is to smell. The sense of smell, though acute in man, is still only a fraction of that of many animals. Man's dependence on sight has developed along with his upright posture, which takes his nose quite a long way away from the scent-laden earth. Animals that depend largely on smell carry their sensitive noses within easy reach of the ground and the scent messages left there. The nose is not the only part of an animal's body that may be used for smelling, however: insects use antennae *(see page 128)*, snakes use their tongues *(see page 143)*. Nor is a nose only used for smelling. Some of the largest and finest noses have other uses, as will be seen.

Smell and taste are known as the chemical senses, in that they detect and identify molecules of chemical substances. As land animals we are used to the idea that smell and the nose are concerned with air, and that taste and the tongue are connected with liquids. But how then can it be said that fishes can smell in water? There is actually a small anatomical difference between the two senses, apart from their separate locations in the nose and mouth. The scent detector cells in the nose are nerve cells directly connected to the brain, while the taste buds of the mouth are skin cells connected to nerve fibres which, in turn, are connected to the brain – in this case the link is not quite so direct. Nonetheless the two senses are very closely connected: when you eat something, its 'flavour' is often constituted largely by its smell – molecules of that substance reach the nasal membranes through the back of the throat.

A nose usually contains two nostrils, opening into nasal passages that lead down to the lungs: a major function of the nose (except in fishes which use gills for breathing) is to channel air into the lungs, filtering, warming and moistening it first. At the same time molecules of chemical vapour fall on the olfactory (smell) membranes lining the olfactory clefts – small but convoluted chambers opening off the main air passages through the nose. The smell receptor cells in this membrane open to its surface with little 'brushes' of waving hair-like cilia, which increase surface area so that there is a much greater chance of smell molecules coming into contact with receptor cells. The sheer number of receptor cells and the total surface area of the membrane is one of the significant factors in comparing the sense of smell of different animals: man has about 5 million cells while a dog may have more than 200 million. When one considers that often only a couple of molecules of scent among the millions of molecules taken in with every breath are needed to react with the receptor cells and be classified by the brain, it is perhaps not really surprising that a dog's sense of smell is so acute. Furthermore, not only is the dog capable of identifying minute concentrations of scent, it can also pick out and follow a single subtle scent from among the hundreds that it continually encounters.

Almost all creatures have some kind of sense of smell, from the invertebrates with no noses but sensory cells on their bodies, to insects with antennae, and fish that smell through their noses in water. Smell is not perhaps so important among most amphibians, but reptiles – especially snakes – make use of it. In mammals the sense of smell is particularly well developed: many mammals could manage to live by smell alone. It is only among the birds that smell does not seem to be much used, although the strange and primitive kiwi hunts nocturnally by smell and has nostrils at the tip, not the base, of its beak. Since this is an area that science has hardly explored, however, it may be that birds use a sense of smell more than we realize.

A sense of smell is extremely useful to animals for a number of purposes. A predator tracks its prey by scent: the prey may also smell a hunter, and escape. Many creatures find their food by smell, like the bats that visit pungent night-flowering plants to suck their nectar. A parasitic wood wasp can smell, through wood, the grub on which she will lay her eggs, providing food for her own young. A mother seal identifies her pup on the beach, among hundreds of other pups, by its smell, although when she is in water her nostrils are closed and her sense of smell cut off. It is well known that a mother sheep will allow a lamb to starve rather than feed it if it has the wrong smell. Animals identify others of

A sheep cleans her newborn lamb to remove all traces of odorous fluid left from its sojourn in her womb. She will also eat the placenta and afterbirth so that there will be no scented matter around whereby a predator (such as a wolf or a fox) could learn of the presence of a newborn lamb nearby. For this reason, newborn mammals often have no scent of their own.

their tribe, and reject strangers, according to their scent. They mark their territory with scent to warn others to keep away; and they leave strong-smelling invitations to sexual encounters.

The biggest noses in nature, however, do not necessarily indicate the best sense of smell – the relative size of the olfactory section to the brain is a better guide, but this is not externally visible. The largest and most versatile nose in creation surely belongs to the elephant; and while its nose, or trunk, accommodates a good sense of smell, it is also modified for a wide range of other duties. This is the nose as a power tool, an amazing amalgam of strength and precision. With its trunk an African elephant can reach 6 m up from ground level; it can pick leaves or fruits and place them in its mouth. It can push, pull, or carry objects as heavy as tree-trunks. It can suck up water for drinking or showering, and spray itself with cooling dust or

The male proboscis monkey's swollen nose is over 7 cm long, and hangs down over its mouth, sometimes almost touching its chin. In the dense forests of Borneo where it lives, calling is the best way of communicating at a distance, and the nose acts as an amplifier when he calls.

An elephant's long trunk not only enables it to drink from pools without having to bend down, but is also able to reach branches well above the elephant's head. The elephant can even stand completely submerged under water, using its trunk as a snorkel, or it can use the trunk as a hose to give itself a cooling shower. Baby elephants have been known to suck their trunks, just as human babies suck their thumbs.

mud. It can breathe through its trunk like a snorkel when swimming. It can protect itself and its young against attack or communicate closely by touching delicately with its sensitive trunk tip. It can communicate at a distance using the trunk as a sound chamber to amplify its thundering call.

A large mobile nose, like that of the elephant, may also be known as a proboscis: hence the name of the proboscis monkey, a curious-looking primate, the male of which has a long bulbous nose that hangs over his mouth. This kind of nose is used as a resonating chamber, and is erected to amplify the male's threatening call through the forest canopy of Borneo.

Another species in which the male sports an exaggerated nose is the elephant seal, largest of all seals at 5–6 m long and up to 3500 kg in weight. The huge, bulging nose of the mature male is used during the breeding season, when the seals gather in vast herds on the shores of California or the South Atlantic islands. The proboscis of the mature male bulges with the combined efforts of blood, muscle and air, and amplifies his defiant bellowing at other males.

The sense of smell is acute among fishes, especially in the shark, which can detect a drop of

This male elephant seal may look as if he is smiling, but he is actually bellowing a warning to other males to keep away from his harem of females. The bright pink lining of his mouth probably serves to reinforce the warning. His inflated nose acts as a resonator for his bellowing calls.

blood at a distance in the turbulent waters of the ocean. The most extraordinary piscine nose, however, must be that of the Atlantic salmon. Born in the headwaters of a rushing highland stream, the salmon leaves for the sea while still only a few centimetres long. Several years later, grown to maturity in the ocean, it navigates its way (perhaps using the sun and stars for guidance) to the coasts of its homeland. Then, of all the rivers running into the sea, it chooses the one that will lead up to the tributaries and eventually the precise stream of its birth, where it will mate and probably die. It is now known that the salmon finds its home stream by its smell; but how it picks the right one unerringly from such distance and after such a long absence is a feat beyond our imagining.

Undaunted by the powerful rush of water, a salmon leaps a waterfall on its journey upstream to spawn, after a journey of maybe 3000 km. An acute sense of smell has enabled it to leave the ocean at the estuary of the very river in which it hatched years before. Local conditions of soil, vegetation, type of stream bottom and general geology give each stream its own peculiar odour.

Mouths and tongues

The obvious function of a mouth is to admit food into the body of an animal. But that, needless to say, is not the whole story. Food has not only to be admitted: it must be held still if alive, tasted, bitten or squashed into a suitable shape, lubricated with saliva from skin glands in the mouth, and swallowed. In many cases the mouth is also the weapon used to catch the food: the beaks of birds *(see page 150)* and the mouthparts of insects *(see page 158)* are often especially adapted for this purpose.

Mouths may also be responsible for protection, defence, communication, and other jobs not directly concerned with food and eating. The mouthbrooder cichlid, for example, is a fish that uses its mouth as a nest for its eggs. The female cichlid lays her eggs, the male sheds his fertilizing sperm over them, and the female then takes them into her mouth to protect and conceal them until they hatch. Mouths are involved with communication, by facial expression and by sound: by amplifying and changing the pitch of sounds produced by the vocal apparatus in the throat, the mouth converts noise into recognizable vocal signals, even – in the case of human beings – into speech. The general rule among animals is that the male uses his larger mouth and louder vocalization to a greater extent than the female, usually to proclaim his superiority, territory, and intractability towards other males.

For the most part, though, mouths are primarily concerned with food: and the shape, size, position and design of an animal's mouth is a good indication of its preferred diet. Different animals may have the same type of mouth to deal with a certain kind of diet, while other very similar animals may have quite different mouths. For example, on the savannahs of Africa (or in the paddocks of a zoo) the black rhinoceros and the white rhinoceros are both grey in colour and not easy to distinguish from each other at a distance. Closer up, however, it is evident that the black rhino has a pointed, very mobile upper lip (its name

The golden cat's mouth is a typical mammalian carnivore mouth. It is surrounded by sensory hairs, in particular the extremely long whiskers characteristic of the cat family. The jaws are well armed with teeth of various kinds, each adapted to the meat diet. The canines or fangs are long and backward-curving, and overlap when the mouth closes to give a vice-like grip on the prey. The cat uses its mouth to kill, sinking its canines into the prey's throat. The incisors at the front are used as scrapers to get small pieces of flesh off the bones, and the massive carnassial teeth at the side act like shears to slice off flesh and crack bones. Still further back are the molars for grinding the food into small enough pieces to swallow. The tongue is well equipped with taste buds.

Mouths tend to be adapted to suit the diet of their owners. The copperband butterfly fish *(right)* lives in coral reefs and has a long tubular snout for probing into nooks and crannies in the reef in search of food, armed with fine bristle-like teeth for delicately extracting small animals from the bottom of crevices.

(Below left) The horse is a grazing animal, and has a soft tender muzzle and rather square lips well supplied with sensory hairs. Like all animal vegetarians, its teeth grow continuously as they are worn down on top. Between the front and back teeth the horse has a large gap where the tongue mixes the grass with saliva.

(Below right) Another carnivore, the dog has teeth which are very similar to the cat's. The dog has much longer jaws than the cat, and uses its teeth for attacking as well as for killing its prey, since it lacks slashing claws.

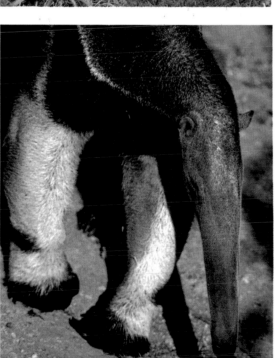

does not refer to its colour but is derived from the Afrikaaner word for 'wide' referring to its mouth). Both rhinos are browsing plant eaters, but the pointed prehensile lip of the black is ideal for hooking round shrub leaves, while the flat lip of the white allows it to graze grass.

Similar diets lead to similar mouths on animals which may otherwise be quite different. This is an example of what scientists call 'convergent evolution' – arriving at the same design by different paths. Animals like the anteater and the spiny echidna, for example, have long narrow muzzles and small mouths, for poking into narrow crevices in the nests of their insect prey: their tongues are long and sticky in order to lap up the insects. American scientists have also suggested recently that the mouths of the baleen whales and the flamingo are examples of convergent evoluton, having evolved as the most efficient shape for their filter feeding habits *(diagram a)*. The whale uses its

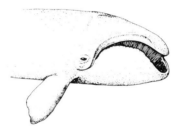

(a) Both the baleen whale *(left)* and the flamingo *(below)* are filter feeders, straining water through sieve-like plates to extract the tiny plants and animals suspended in it. Water is taken into the mouth, then forced out through the filter, leaving its suspended food on the sieve-plates for the tongue to lick off.

vast mouth to take in a great gulp of sea water containing thousands of tiny shrimps or 'krill' *(page 65)*. Then it pushes its tongue – which, in the blue whale, may weigh as much as an elephant – against the inside of the baleen plates, ejecting the water and leaving a porridge of crustacea inside to be swallowed. The flamingo feeds in a similar fashion, hanging its oddly-shaped beak upside-

The tamandua *(above left)* and the giant anteater *(left)* both live on a diet of termites and ants. Both have very long snouts for probing into termite nests. The nostrils are at the tip of the snout, which is held close to the ground to pick up the scent of the prey as the anteater ambles along. These animals have no teeth, but catch their prey on a sticky worm-like tongue that may be 25 cm long – this can reach a long way into a termite nest.

Filter feeding is obviously a very successful mode of life: some flamingo populations are measured in millions rather than thousands. Their filtering system is so efficient that each bird can consume about one tenth of its body weight every day. The flamingo's strangely angled beak is actually upside-down while it is feeding, and the nostrils are high on the beak to allow it to breathe while filtering. Situated at the edge of each half of the beak and visible as a stripey line in this picture, the filter plates are designed so that they exactly interlock when the two halves of the beak are brought together. As the bird feeds, it takes in water, then closes its mouth. The large fleshy tongue lies in a deep groove in the floor of the lower bill (uppermost when feeding), and acts like a piston, pushing the water out through the sieve. The surface of the tongue is armed with little hooks for scraping the food off the filter plates.

An alligator snapping turtle lies in wait for a passing fish, well camouflaged against the muddy river bed. Like all turtles and tortoises, it has no teeth, but its jaws are covered in a sharp-edged horny beak suitable for shearing flesh. On the floor of its mouth is a fleshy pink worm-like lure, which the turtle waggles to attract fish. Eager to seize the 'worm', a fish may swim right in the turtle's gaping mouth.

(*Below*) A western coachwhip snake flickers its forked tongue. Surprisingly, this tongue is involved in the snake's olfactory process. In the roof of the mouth is a pair of pits lined with smell sensors, the Jacobson's organs. All snakes and lizards having these organs have forked flickering tongues. It is thought that these waft airborne chemicals towards the Jacobson's organs.

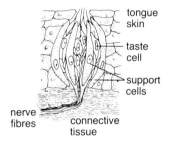

(b) Section through the tongue skin of a vertebrate showing a taste bud.

down in fresh or brackish water and filtering off the small organisms gulped into its beak.

The mouth of most vertebrate animals contains a tongue, an extremely flexible muscle which is charged, among other things, with the vital duty of tasting. The taste of its food may be the only grounds on which a creature can judge whether the food is positively likeable, merely edible, or even poisonous – a judgement that must be made quickly. The sense of taste lies in 'taste buds', cells in the surface of the tongue skin *(diagram b)* that can detect dissolved chemicals and send a message via nerve fibres to the brain, which identifies the 'taste'. It is said that people can detect four basic tastes: sweet, salty, sour, and bitter, being most sensitive to the last. Other animals evidently cannot taste all of those. The amount of taste buds on an animal's tongue will obviously render it more or less sensitive to taste, but the ratio does not seem to be proportionate. A bird with only 25 tastebuds on its tongue can still detect and spit out a bitter poisonous caterpillar; while a cow, with 25 000 tastebuds, rarely uses them on anything other than grass. Given our taste for a diet of varied flavours, it is a sad fact that the tastebuds of humans decline rapidly in number from 10 000 at birth to a mere

3000 in old age. The food of our childhood did not really taste better – but we tasted it more accurately.

The distinctly forked tongues of snakes and lizards have a special function – they 'taste' the air, performing a smelling function that is usually the responsibility of the nose. The snake's tongue detects airborne chemicals rather than dissolved ones. It does this with the help of a special sense organ, called Jacobsen's organ, which is located in the roof of its mouth. The snake constantly flickers its tongue in and out, each time carrying scent molecules from its surroundings to the Jacobsen's organ where the molecules are 'recognized' by nerve cells connected to the brain.

Certain lizards do not have forked tongues. The chameleon's is long, fleshy and sticky, with a bulbous tip. It can be projected from the mouth to a distance greater than its own body length, and almost faster than the eye can see. The tongue is a predatory trap. The chameleon detects its prey – insects or small animals – with its strange eyes, and it creeps slowly within range. Then there is an instantaneous flash of tongue, and the prey disappears, borne on the sticky tip of the chameleon's tongue into its gaping mouth.

A chameleon catches a fly on its sticky tongue. It is using its prehensile tail as an anchor while stretching out towards the flower. This tongue can stretch out to almost twice the chameleon's body length, and when not in use, it is folded around a long thin bone in the floor of the mouth. When the chameleon strikes at its prey, this bone is levered forwards and the muscles contract to extend the tongue. This accounts for the kink in the tongue seen in this picture.

Teeth

The smile on the face of the crocodile is given a treacherous guile by its fringe of numerous pointed teeth. The alligator is distinguished by its broader snout, and when its mouth is closed only the teeth in the lower jaw are visible. The crocodile, however, does not conceal its arsenal of teeth, including a particularly menacing fourth pair in the lower jaw. One look at a crocodile is enough to know that this is no harmless herbivore. When teeth are on display, it is usually for the good reason that they are weapons, and often extremely daunting from the point of view of the prospective prey.

Teeth belong in the faces of vertebrates. They are often pronounced among the fishes; insignificant, or absent, among the amphibians; distinctive and varied among reptiles; totally absent in the living bird species; and again, in the mammals, well developed and widely adapted for different diets. Teeth are used mainly to chop or grind food small enough to be swallowed; but when it comes to predators, teeth are also used to trap, injure and kill. The study of an animal's dental equipment can usually tell a scientist what kind of food the animal eats and hence reveals much of its way of life. Since teeth are among the most durable of animal parts, they can tell us about the lives of animals that no longer exist, like the ferocious reptilian dinosaurs and the flamboyantly-tusked mammoths. Teeth may be as personal as fingerprints, betraying the identity, age, and physical condition of their owner. Many a corpse has been identified by its teeth.

A typical tooth *(diagram a)* consists of a peg of very hard bone-like material called dentine, or ivory. The outer surfaces are covered with a fine layer of even harder, crystalline, enamel. Within the tooth there is a soft pulpy core richly endowed with blood vessels and nerve endings. The tooth may be fixed into the mouth in several ways. It may be slotted into a socket in the jawbone and fixed there with a substance called cement; this is the general pattern in mammalian dentition. Teeth may also grow in the skin and constantly move round the jaw into a position for use, as in the sharks. Alternatively they may just be stuck on the top of the jawbone with a hard tissue, as in many fishes.

Teeth come in different shapes for many different purposes. Where the top or crown of the tooth is broad and flat, it usually indicates that the tooth is

(a) A section through a single mammalian tooth, showing how it is embedded in solid jaw bone. Blood vessels and nerves run to the pulpy centre, supplying food and oxygen and conveying information to the brain about the food the animal is eating. A white cap of enamel coats the exposed part of the tooth, and the dentine is shown by shading.

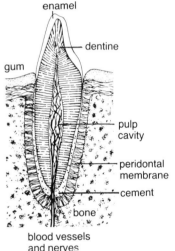

Like the teeth of many fish, alligator teeth are all the same shape, but unlike those of fish, they are firmly embedded in deep sockets in the alligator's jaw bones. In structure they resemble those of mammals. As each tooth wears down, it is replaced by a new tooth which starts growing in a cavity at its base until eventually it pushes out the existing tooth. This is useful, since American alligators may live for over 50 years. Note how the teeth interlock – this helps the alligator grip a struggling prey animal.

A young American crocodile displays a mouth full of teeth. These sharp pointed teeth can easily penetrate flesh to hold and kill the prey, and are also capable of cutting it to pieces, but they are not suitable for chewing, so chunks of flesh have to be swallowed whole. When its mouth is shut, the large teeth on the lower jaw will not be visible, as they fit into the notches between the upper teeth.

used for grinding – probably tough plant food. Plants, being full of indigestible fibres, must be broken down as much as possible even before being passed on in the body to the destructive stomach acids. Grazing animals or ruminants, such as cows, semi-digest chewed vegetation for a while before regurgitating it and grinding it up for a second time with their big, flat cheek teeth, or molars. Rodents (rats, mice and their relatives) also have grinding molars.

Teeth with straight, sharp-edged crowns, like our front teeth or 'incisors', are used for clipping. Herbivores such as rabbits have clipping teeth at the front of the mouth to cut vegetation, and flat molars at the back of the mouth for grinding the food into a mush. The big incisors of a beaver make short work of wood – not the sort of thing human beings can get their teeth into.

Large, pointed, conical teeth are not much good for cutting up food, but they can be sunk into wriggling prey to prevent them from escaping, and they can also tear a quick entry into the tough hide of a freshly-killed victim. When they grow at the front four corners of the mouth, these teeth are known as canines or 'dog teeth', and they are pronounced among mammalian predators such as dogs and wolves, bears, and the big wild cats including lions and tigers.

Rows of broad teeth with sharper edges and a more jagged outline serve as secondary teeth for carnivores, to crunch up flesh and sinew into chunks small enough to be swallowed; similar teeth may be used by omnivores, animals with a mixed diet, such as many of the monkeys, and ourselves.

Triangular teeth with finely serrated sharp edges will slash and cut as cleanly as a razor, though sometimes with the force of a wielded axe behind

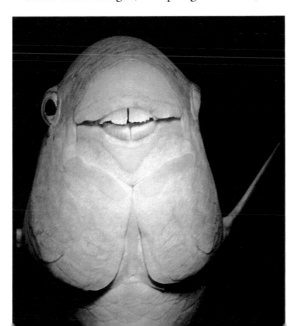

The parrot fish feeds on algae and coral from coral reefs, and is one of the main causes of sand production from coral reefs. Its teeth are fused together to form a beak-like edge to the jaws for scraping at the coral, and it also has large flat-topped grinding teeth at the back of its throat for crushing the coral and algal mixture. The indigestible sand is excreted.

145

Fish teeth are thought to have evolved from scales. In cartilaginous fish, such as sharks and rays, the teeth develop from the skin, and move forward as they grow. In bony fish they are found on the jaws and other bones of the mouth, the palate and even on the tongue in some species. These teeth are relatively simple and unspecialized, with sharp points for penetrating the prey to trap it, and serrated edges for cutting and slicing the flesh to swallow it. In bony fish they are made of dentine, with a harder cover of vitro-dentine. (*Below*) A tooth of the extinct giant shark, *Carcharodon*.

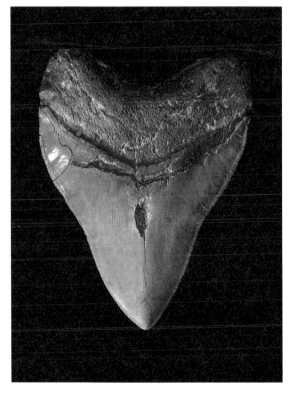

them. Such is the dentistry of the great white shark and other large sharks. They can cut off the limb of an unfortunate swimmer with such quick, surgical efficiency that the victim's only immediate sensation may be of a hard knock. The teeth of the great white shark, each up to 7.5 cm long, are arranged in concentric rows, and have attained a popular notoriety through the cinema in recent years; but even more menacing are the teeth of its extinct ancestor, the giant shark – the same design, but about three times larger. This giant shark *Carcharodon* reached 27 m in length. Had it existed in human times, it had jaws large enough to swallow several fully-grown men without even bothering to cut them in half first. All sharks have skin covered in scales which are really tiny teeth.

Perhaps the most remarkable adaptation of teeth is tusks; and the most remarkable tusks belong to the elephants. The tusks are modified upper incisors, and contain – like other teeth – a pulpy sensitive core which, if exposed by damage, causes considerable pain. The ivory grows continuously throughout the animal's lifespan – some sixty to seventy years at most – but the tips of the tusks are constantly worn down or chipped, so they rarely

A western diamondback rattlesnake strikes at an intruder. The snake's jaws are specially hinged to allow it to open them extremely wide. This is necessary because the fangs curve inwards and need to be plunged vertically into the prey. When not in use, they are folded back against the roof of the mouth *(see diagram)*. The snake's windpipe is protruding at the bottom of its mouth – this is so that the snake can still breathe after it has a mouthful of prey. Snake's teeth are not designed for chewing or grinding, so the prey has to be swallowed whole.

fangs folded back

poison sac

fangs hinged forward to strike

(b) Attack mechanism of a venomous snake

(Far right) Drops of venom can be seen trickling out from the hollow grooves in each fang. The amount of venom in one pair of these fangs may contain enough poison to kill 45 people.

natural death in old age, for once its last teeth become worn and useless it inevitably starves to death.

Another unique adaptation of the upper incisors occurs among poisonous snakes – the fang. Apart from small sharp teeth in its jaws, a venomous snake such as the rattlesnake has two long curved fangs (modified incisors) at the front of its mouth. When the snake opens it mouth to strike at prey, the hinged jaw bones act like a series of levers to push the fangs forward into a vertical position *(diagram b)*. As the snake actually strikes, poison from the venom gland (a modified sweat gland) in the head passes into a hollow groove in each fang, and is neatly injected into the prey. The rest of the snake's teeth are only used to hold the meal in position while it is being swallowed: the snake does not chew or grind its food but swallows it whole. Large meals and slow digestion mean that snakes only need to eat every few weeks.

Beaks

The shape of a bird's beak usually gives an accurate indication of its preferred diet. In some birds, however, such as the puffin, with it broad striped beak, unlike that of any other seabird, or the shoebill stork, with its huge flattened bill – a direct contrast with the rapier-pointed beaks of other storks – the bill does not seem to be directly adapted to the food taken. It is a zoological brain-teaser to try to work out why such beaks might have evolved. Like feathers, beaks are unique to birds, and compensate admirably for the teeth that all modern birds have dispensed with. A major reason for this toothlessness may be that when birds took to the air, their bodies became adapted to put lightness before all other design demands, and a beak is much lighter than teeth and heavy jaw muscles would be.

The beak of a bird is derived from skin, the three layers of epidermal cells overlying its jaws. The innermost layer is the germinating layer, where the cells multiply; the second is the granular layer, where the tough protein keratin is formed; and the horny outer layer consists of very tough, dead, keratinized cells. The beak is continually growing to make up for the heavy wear and tear it suffers in normal usage. Caged birds often require special hard food supplements to keep their beaks in trim, and the beaks may even need to be clipped if they are not used enough.

Generally the beak tip and cutting edges are as hard and insensitive as our fingernails, but in some birds there are many tiny nerve cells in the beak. The bills of most ducks, for example, are softer than those of other birds and are well supplied with these tactile cells, so that the duck can feel for worms or other food in soft mud or cloudy water where it cannot see. Wading birds like the snipe and curlew also have tactile cells at the tip of their long narrow beaks, with which they probe damp mud or sand to feel for the small worms and shellfish that burrow out of sight.

Many species of birds live largely or entirely on fish, and they tend to have long, strong, dagger-like beaks with which to spear their prey. The gannet is a coastal bird that flies over the sea searching for fish, then plummets into the water to spear them. Its beak is a similar shape to that of the kingfisher, which hunts in the same way over fresh water; but

These puffins sport beaks that look as if they have been painted for a carnival. The exact purpose of the colouring is not clear, but it is much duller out of the breeding season when the outer sheath of the bill is shed. In young birds the bill is black. The puffin feeds on fish, which it catches while diving. It is able to carry as many as 20 fish in its beak by holding them with its tongue against the serrated upper mandible.

The shoebill stork or whalebird lives in the papyrus swamps of the Upper White Nile and East Africa. Its main food is lungfish, for which it probes in the mud with its grotesque hooked bill, which may be up to 20 cm long. The bill is also used as a scoop to catch frogs and other small vertebrates. The shoebill expresses its emotions by 'clappering' its bill, pointing it skywards and making a noise like an old-fashioned motorcycle.

151

A flash of iridescent blue wings announces
the arrival of a kingfisher at its nest hole,
bringing a fish for its hungry family. Inside
the burrow, the nestlings will line up in an
orderly queue so that each one is fed in
turn as the parents arrive with their
catches. To catch its prey, the kingfisher
dives from its favourite perch to catch the
fish underwater, seizing it in its sharp-edged
beak and taking it back to its perch to be
killed. If it were going to eat the fish itself,
the kingfisher would first turn it around in
its bill in order to swallow it headfirst, since
it swallows fish whole. This may take quite
a lot of juggling.

Some beaks are clearly adapted for a particular function, while others are inexplicably bizarre. The kingfisher (*far left*) has a long pointed beak which it uses to catch fish, opening its beak as it chases them underwater. The skimmer (*below*) has a unique fishing technique. It hunts at dusk or by night, flying low across the water, opening its beak and trailing its lower mandible in the water as it flies. This creates a line of light in its wake, which attracts fish. The bird then returns along the same path to pick up the fish, its beak snapping shut on contact with an edible object. To compensate for the wear caused by friction with the water, the skimmer's lower mandible grows faster than the upper one, and in adult birds it is usually longer. The bill is laterally flattened, which reduces drag to a minimum.

The toucan's bill (*left*) is the subject of much controversy – no-one really knows why it is so large. The toucan feeds mainly on fruit, and one suggestion is that the long bill would help it reach fruit borne on branches too slender to take the bird's weight. It is certainly a remarkable piece of engineering, much lighter than it looks; under the horny cover is a honeycomb of fibres, conferring strength without much weight. It has serrated edges and a pincer-like tip with which the bird can manipulate food. But to transfer food to the throat requires it to be thrown in the air and caught in the base of the bill.

The vulture *(right)* and the American bald eagle *(below)* both have powerful hooked bills with sharp edges for tearing at flesh. The eagle is a hunter, using its large talons to seize its prey, while the vulture is a scavenger, feeding on dead carcasses. Even its tongue is rough for rasping flesh from bones.

perhaps because of the depth and speed of the gannet's dive, its beak has no external nostrils into which water might be forced. The skimmer is another fishing bird with a sharply pointed beak of unique design. It flies low over the surface of the water with its mouth open, the lower mandible cleaving the surface to pick up any small fish or shrimps. Because of this feeding habit the lower mandible is worn down by water friction, and consequently it grows faster than the upper mandible. If the skimmer is prevented from skimming the water, as in captivity, its lower mandible will grow to twice the length of the upper, and is virtually useless for any other method of feeding.

The beaks of birds of prey, such as eagles, are almost always very strong and hooked, with a sharp pointed tip for holding wriggling prey animals. Vultures, even though they rely mostly on the dead prey killed by other animals, have a similar and even larger beak: ripping apart bones and sinews of dead mammals is tough work.

Some of the largest beaks in the bird world belong to the hornbills and toucans, birds of the tropical forest. They appear to be clumsy and heavy, but the

Storks are highly successful fishermen, and have been around for at least 50 million years. Only now are their numbers seriously threatened as their wetland habitats are drained or polluted. The black stork is found across Europe and Asia from Spain to China, and is a rather shy bird, nesting on cliffs or in tall forest trees. The stork seeks its prey while standing in water, and seizes it with the long pointed beak. Its eyes are directed slightly forward, giving it a certain degree of binocular vision, and it seems well able to allow for the effect of light refraction at the water surface when lining up its beak on the fish. It is thought that the relative angle of its eyes and beak assist in this.

(*Right*) Like the black stork, the great white egret is a skilled fisherman. Mirrored in the calm water, it scans the shallows for fish.

(*Far right*) The pink-backed pelican uses its distendable throat pouch as a fishing net, scooping fish and crustaceans from the water as it swims. When it opens its bill underwater, the sudden inflow of water carries the prey in with it. Then the pelican raises its head to drain out the water before swallowing the prey. Its pouch can hold considerably more than its stomach but, contrary to popular belief, is not used for storing food.

The roseate spoonbill has a slightly specialized bill. As it feeds, it sweeps its partly open bill from side to side, filtering crustaceans from the water.

interior material of the bill – and the horny knob or 'casque' that may top it in some hornbills – is spongy and light in structure. However it is not always immediately· clear why these birds have beaks of such size. The toucans live mainly on fruit, and can operate their brightly-coloured beaks with surprising delicacy when peeling and penetrating fruits with tough or spiny skins. Hornbills include poisonous creatures like snakes and scorpions within their diet. They render their prey harmless by tossing, biting and thoroughly crushing it in their long beaks before eating it.

Another type of beak adaptation is evident in the birds that feed on flower nectar, such as hummingbirds and sunbirds. Hummingbird beaks are long and very slim, forming a proboscis through which the bird can extend its long tongue to suck up nectar while hovering for a few seconds in front of the flower. The longest beak belongs to the sword-billed hummingbird, and measures as much as the bird itself from head to tail. This species takes its nectar from flowers with long tube-shaped corollas; other hummingbirds have curved beaks for feeding at curved flowers. Such flowers often rely on the hummingbirds for their pollination, suggesting that the relationship between the bird and the plant may be so close that the two different life forms must have evolved together.

157

Insect mouthparts

A silk moth caterpillar spends most ot its time eating. Having started its life in an egg already laid on its food plant, it has little need for travel. It has a heavy pair of serrated mandibles for cutting into leaf tissue, and small jointed palps to help guide the leaf edge between the mandibles. Just behind this are a small cluster of ocelli – simple eyes that do little more than register light and dark. The caterpillar's simple body plan is designed just for eating and growing. The complex sensory organs and highly co-ordinated flight apparatus of the adult moth are needed only in order to seek a mate, and in doing so disperse the species over a wider area. Until the insect has accumulated enough body tissue to form the adult, what better plan than a simple bag, whose skin can be shed for a larger one at intervals.

(Far right) An '89' butterfly rests, its proboscis neatly rolled up, and its two palps pointing upwards. Only the maxillae form the long proboscis, in the form of two half-tubes fitted together to form a tube through which the butterfly sucks nectar from flowers. This is its only food – an energy-rich sugary fluid to power the flight process. When feeding, the proboscis unrolls to provide a long drinking straw capable of reaching deep into flowers.

(a) Head of a cockroach showing typical insect biting mouthparts.

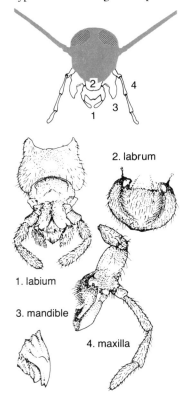

2. labrum

1. labium

3. mandible

4. maxilla

The powerful mandibles of the hornet are not used for feeding: it uses its 'tongue' to lap up juices from flowers and fruits. The mandibles are used to cut and chew wood to make its nest. Wood pulp is mixed with saliva and converted by the hornet into a paper substance from which an elaborate tiered nest is made which may house thousands of individuals.

A caterpillar straddles the rim of a leaf and its jaws, like tiny secateurs, clip away neat semicircular holes and erode the leaf at a prodigious speed. A couple of months later, a butterfly pauses briefly on a flower and uncurls a long 'tongue' or proboscis with which it probes the heart of the bloom to suck up nectar. The butterfly was once the caterpillar, but since its metamorphosis it has adopted a completely different diet, and consequently its mouthparts have had to change shape dramatically. The mouthparts of both butterfly and caterpillar, however, are formed from the same basic pattern, a pattern shared by all insects. Just as birds' beaks are adapted to their eating habits, so too are insect mouthparts.

To look at – and it may involve peering through a microscope – insects' mouths are nothing like ours. They are composed of several parts clustered around the tiny opening through which the food enters the body. A cockroach is an example of an insect with typical mouthparts *(diagram a)* all constructed from the body cuticle. On the front of the face is a hard flap that corresponds to our upper lip, called the labrum. Just behind it, one on each side of the mouth opening, are two mandibles, the cutting jaws of the insect. They often look, and act, like clippers, and in some insects they are strong enough to cut through wood or even metal.

Behind the mandibles is another pair of jaw-like structures, the maxillae. These may be simple in shape but often they bear soft lip-like appendages, and projections like tiny antennae, called palps. These bear many sensilla *(see page 128)*, sensitive cells for tasting, smelling, and touching the food. The maxillae are not usually designed for cutting or chewing food, but they may be used to hold it steady and pass it forwards through the chopping mandibles. Finally at the back of the mouth is the 'lower

159

lip' or labium, which may also carry palps with sensory cells.

An insect with a most unusual labium is the dragonfly larva, a predatory denizen of fresh-water ponds and streams. It has a long, hinged labium which is folded away under its head when at rest; but when a prey animal comes within reach, the labium, suddenly pumped full of blood, shoots forward to trap the prey and pull it back to be demolished by the mandibles. It is such an effective weapon that the voracious dragonfly larva thinks nothing of taking on prey as large or larger than itself, such as small fishes or baby frogs.

By complete contrast, some insects have no mouthparts at all. The short life of an adult male moth, for example, may be concerned exclusively with finding a mate and reproducing; and as feeding would be a waste of precious time it dispenses with mouthparts completely, and never feeds.

Like all wasps, the hunting wasp *Ammophila* feeds on liquid food, sucking it up with its long 'tongue'. Note how it is investigating the food with its slender jointed palps. These are laden with sensillae to ascertain the edibility of the food.

(Below) A dragonfly nymph feeds on a water louse firmly clasped in its mask, the large flat structure hanging below its head. The mask, which is really a modified labium, is kept folded under the nymph as it lies in wait for passing water creatures, then shot out by hydraulic pressure to impale the pray on its vicious claws.

A leaf cutter ant struggles with the large piece of leaf it has just cut off, trying to grip it with its mandibles in such a position that it can be carried back to the nest.

(*Below*) This massive pair of mandibles belongs to a South American harlequin beetle, and is probably used only once in its life. The harlequin beetle larva lives inside a tree, where it feeds on the wood. Here it pupates, and when the beetle emerges, it has to chew its way to freedom. The adult beetle probably does not feed at all, its sole purpose being to reproduce.

Birds have perfected vocal communications and obviously have good hearing. In many cases it is so good that it shows up the inadequacies of our own ears, for a majority of birds can hear sounds ten times softer than we can. Birds cannot hear frequencies above 10 to 12 Khz – only mammals can, among vertebrates. We can hear 15–20 Khz depending on age. Sharp as they are, the ears of birds are all but invisible, consisting usually of small openings tucked under a tuft of feathers. Only some of the owls, such as the eagle owl, have what seem to be ears – tufts of feathers sticking up from the top of the head – while the facial feathers of the barn owl act as horns to direct the sound into the ears.

Underneath the head feathers, owls have relatively large ear-holes with fleshy flaps around them, which help to channel sound waves into the ear. In many owl species the ears are asymmetrical: they are of different sizes and set at different levels on either side of the owl's head. Since the head is wide and the ears therefore set far apart, a noise coming from an angle will reach one ear a fraction of a second earlier than the other, and it is this infinitesimal delay that indicates the direction of the sound source. The owl reacts by turning its

Many desert animals have large ears, and the jack rabbit is no exception. It has been suggested that large ears, with their network·of blood vessels, may serve to radiate heat to the sky while the animal is resting in the shade, so helping to lower its body temperature.

The 'ears' of this long-eared owl are not ears at all, but merely decorative feathers. The owl relies mainly on its hearing when hunting. Its ears are placed asymmetrically on either side of its head, and the sounds received by them are interpreted to give a very accurate three-dimensional sound map of its surroundings.

scale – some have ears, while others have none. Crocodiles and alligators, for example, have external ear openings which they can close when in water. Snakes have no ears and cannot hear, but they are very sensitive to vibrations in the ground, which they feel through their bodies. Since a rattlesnake cannot hear its own rattle, nor any snake hear another snake hiss, one wonders how they first learned to make and use sounds as warning of their presence. A tortoise is as quiet as a snake, but it has ear openings and quite keen hearing.

The bat-eared fox of the African deserts looks more like a small jackal. Its ears like those of the jack rabbit, are thought to radiate heat back to the atmosphere. Its warm reddish coloration affords good camouflage against the red desert soil of its habitat.

(*Below*) Another desert animal with large ears, the Hotson's jerboa from Iran is thoroughly adapted to a desert life. It has no sweat glands, as these would loose too much vital water. Instead, it excretes only a very concentrated urine, and has other cooling devices, such as the large ears.

(*Overleaf*) The ears of an elephant seem out of proportion even on an animal of its size. They are thought to act as cooling devices: they have a large surface area and are well supplied with blood vessels, so if the animal stands in the shade, heat can easily be lost from them by convection and radiation. An angry elephant will also use its ears to frighten its opponent. By spreading them out it makes itself look even bigger than it is.

The situtunga lives in African marshes and feeds mainly at dawn and dusk, when it is less conspicuous to predators. At this time of day, hearing is important for the situtunga to detect predators, and it can orientate its ears towards the source of a sound to hear it more acutely.

head (sometimes nearly all the way round) until the sound reaches both ears at exactly the same moment: it is then facing the sound source – such as a squeaking mouse – directly, and can bring its acute eyesight into play as well. The vision is not always needed, however, for the hearing of most owls is so acute that they can catch a small mouse in a totally dark room. The assymmetrical ears help the owl to track moving prey as it hears the sounds at a different volume in either ear. Owls are most sensitive to high-pitched sounds since it is on these wavelengths that their rodent prey transmit their calls. The calls of the owls themselves, however, are generally low-pitched to carry for long distances. As an extra, if indirect, hearing-aid the feathers of owls are very soft and thick, muffling the sounds of their own flight so that the faint sounds of their prey can be perceived more clearly and, in turn, so that the prey is not alerted to their presence.

It is only among mammals that ears become noticeable, even striking, because of the visible external ear-flaps behind the narrow opening of the outer ear tube. The ear is a complex mechanism. Sound waves enter the outer ear and produce vibrations in a taut membrane, the eardrum. The vibrations are transmitted through the middle ear by minute bones to the inner ear, which is a knot of convoluted channels containing ultra-sensitive hairs. Every tiny alteration in the position of these hairs is related to the brain via nerves, and interpreted. The external ear-flaps serve a number of purposes, not necessarily all concerned with hearing. In some mammals they are protective, flopping over the ear opening to prevent the admission of dirt or insects into the delicate ear channels. Among aquatic mammals such as the otter, the ear-flap can be pressed down to close the ear to water.

The most obvious use of the ear-flap, though not necessarily the most important, is to gather and concentrate sound waves. To a certain extent our own ear-flaps do this, but we can improve their

function by cupping a hand behind the ear. It is easy to assume that large ears means acute hearing. But it is often so, as among the bats. Almost all bat species except most fruit bats (which eat by day and seek out their food visually) use sonar or echo-location, sending out of their mouths or nostrils a stream of high-pitched sound waves that bounce off objects and are then received by the bat's large ears. The slightest change of pitch, timing and volume in each ear tells the bat what is near it, whether it be the roof of a dark cave or a moth flying on a moonless night. Among birds, cave-dwelling swifts and oilbirds also use echo-location, but they manage without large ear-flaps.

A significant function of large ear-flaps is heat loss. In the thin skin of the ears, blood passes close to the outer surface of the body and easily loses heat to the outside air. This is evident among rabbits, for example; and those from warmer climates have

larger ears (for greater heat loss) than those from cold climates. Elephants, too, have huge ears relative to body size, and they are also used for temperature regulation. It has been shown that blood passing through the ears of an African elephant may lose as much as 9°C of heat, a valuable cooling device in the hot African summer. To increase the effect, elephants spray their ears with water, and flap them to create cooling air currents.

Many mammals – other than man – can flap their ears; they may be able to lift or lower them or turn them around. This is not only extremely useful for picking up faint sounds and locating their source precisely, but also for communication. Mammals such as horses, cats and dogs will turn their ears forward in amicable attention, but lay them flat against the head when expressing aggression or fear.

An Arctic fox, well insulated against the cold by its thick furry coat, rests in the snow. Compared to the bat-eared fox, or even the red fox, its ears are small. In the Arctic climate where it lives, it cannot afford to lose any more heat than necessary, so its ears – the least furry parts – are small to minimize heat loss.

A false vampire bat swoops in on an unfortunate mouse. Carnivorous bats hunt by night, relying on a sophisticated system of echo location to locate their prey and to avoid obstacles when flying in the dark. The bat emits ultrasonic sound waves outside the range of the human ear. Using its highly developed ears, it measures the delay between the emission of the sound and the return of its echo from a solid object. Its prominent nose leaf, a fold of skin standing up over its nose, is also involved in direction-finding, but its role is not understood.

Hands

A young rhesus monkey is being groomed by its mother. Monkeys, like humans, have thumbs which are arranged more or less opposite the other fingers. This enables them to perform delicate tasks like picking out parasites lodged in the fur, a job for which most mammals have to use their teeth.

We human beings owe our supremacy over the other animals to one feature of physical design above all others: our hands. We can see well, but not as well as an eagle. We can smell, but not as acutely as a dog. We are strong, but no match for a gorilla of the same weight. We can run, but not as fast as a deer. Even with our superior brains, we could well have become extinct long ago were it not for our hands. Hands like ours, with that small miracle of biological engineering – the opposable thumb – have allowed us to extend our abilities by using and making tools. With tools we have accomplished a million biological impossibilities, culminating in the sum of technology that allows a man to walk on the moon. Without the right kind of brain, of course, we could not make tools; nor could we do it without hands.

Hands with thumbs are something that we share with monkeys and apes. At a first glance their hands are very similar to ours, but a second look reveals significant differences. It is evident that the palm area of the apes, such as chimpanzees and gorillas, is much longer than its human equivalent. This means that the fingers are further away from the thumb, which in turn means that they cannot be fully opposed. Opposability is the capacity to make contact between the pad of the thumb and the other four fingers, especially the first, making a very precise pincer grip. Human hands have complete opposability, being able to press the thumb pad fully against any one of the fingertip pads. No other animal can manage that small and apparently simple movement. The primates which come closest to us in opposability, and the detailed precision of movement that it allows, are the baboons and mandrills. They can pick seeds from fruit, or groom each other with great delicacy, although some of this must be attributed to the small size of their hands.

Our hands could only have evolved as they have

A Japanese macaque and her baby are enjoying a meal of fruit while taking a bath in a hot spring. The partially opposable thumb is very useful for gripping and manipulating the fruit.

adapted for tree-swinging seem hardly necessary: but evolution is a slow process and it will take a few million years to change them.

Although the gorilla has the best manual opposability among the apes, it is the chimpanzee which makes the most creative use of its hands. Gorillas and other apes such as orang-utans rarely use tools even if they are readily available, but chimpanzees are inventive in the wild and even more so in captivity where they can learn from

The orang utan also has an opposable thumb, but it is not so long and versatile as that of the chimpanzee. It leads a more arboreal life than the chimpanzee, swinging from tree to tree to move through the forest. So it has very long fingers which can form a strong hook, and long palms, and its feet can also curl round to grip branches.

The hands (and feet) of a chimpanzee are very similar to those of a human, and are also almost naked, allowing for great sensitivity of touch. Perhaps the most characteristic feature is the opposable thumb, which can be rotated in its socket to allow the chimp to grasp objects firmly and to manipulate them.

because we no longer use them for support like feet. This is where the other primates are still at a disadvantage: they use their hands at least partly for locomotion. Many of the primates are brachiators, that is, they move by hanging and swinging from tree branches. Their long palms and strong, calloused fingers are adapted for this method of travel. When bent over, their fingers form a strong hook that can support their whole body if necessary (the hook grip of a spider monkey will even continue to support the monkey after death). The apes can also double-lock their fingers, curling them right into the palm in a way we cannot manage, giving them great prehensile strength. Most of the apes also use their knuckles as 'feet' when they walk on all fours on the ground, and their knuckles are correspondingly large and tough, reducing the possible delicacy of their hand movements. In fact gorillas prefer to move on the ground rather than in the trees, so hands mainly

The squirrel's fingers are much less flexible than those of the apes, but can still be used to manipulate food as it eats. To make up for its less flexible feet, it has sharp claws to help it get a grip on branches.

The sea otter is a user of tools. It lives among kelp beds off the American coast, feeding on sea urchins, shellfish and fish. The first two of these have hard shells, so the sea otter brings up a stone from the sea bed and, floating on its back, it places the stone on its belly. Then it holds the shellfish in its forepaws and smashes it on to the stone to break its shell. Sea otters often have favourite stones, and bring up the same stone again and again. The use of such an anvil calls for considerable manual dexterity, yet the sea otter has only short stubby fingers.

humans. Putting hands to good use is subject to intelligence and necessity. Hands with opposable thumbs are not a prerequisite of tool use, of course. Several birds use objects as tools: the Galapagos finch uses a pointed twig to poke insects out of tree bark, and the thrush uses a stone as an anvil on which to break open snail shells. The sea otter will float on its back while smashing shell-fish against a rock balanced on its chest. Plenty of mammals, especially rodents such as squirrels, will use their forepaws like hands to hold things, or bring food to their mouths. But neither the otter nor the squirrel can be said to have hands. What makes a forepaw into a hand is its opposable thumb; and the ability to use one hand in a prehensile manner independently of the other hand. The otter and the squirrel must use both paws for holding, and can never

obtain a precision grip. The only mammal that uses a forepaw independently to hold its food is the giant panda, which indeed has something that looks like a thumb. But it isn't; it is an oddly enlarged and slightly movable wrist bone, and it is certainly not opposable. Hands have developed to feel – they are the major, and very sensitive, organs of touch; to hold, pull and push, climb and carry; to catch food, peel or dismember it, and convey it to the mouth; to scratch, pick, and explore where the eyes cannot see. One special hand function which is almost uniquely human is the use of the hands in communications. How many people conduct a conversation without making the smallest gesture? Perhaps this has come about because evolution – in enabling man to stand upright – has completely freed our hands from locomotory duty.

175

Feet

Feet are characteristic of land animals in particular. Almost all aerial and many aquatic creatures, however, must also move at least occasionally on land, so they need legs and feet. Limbs developed when the first creatures left the sea and abandoned the support of water for their bodies: they needed something to support their bodies on land, and the vertebrate skeleton with four limbs was the result (*see page 36*).

All vertebrate limbs, like our own arms and legs, are based on the same bone structure, and the feet of all vertebrates except the fishes contain the same provision for five toes or digits. Some animals have lost or reduced the number of digits during their evolution: for example, hippos have four toes, rhinos have three, cattle have two, horses only one. But they all started with the same basic bones. The same is true of birds, but they adapted their two pairs of limbs in radically different ways. The forelimbs became wings and the hind limbs developed into specialized legs and feet. The reptiles and amphibians have generally retained the five-toed pattern, except for the snakes which dispensed with limbs and feet altogether. Some mammals, in returning to a marine life, adapted their five-toed

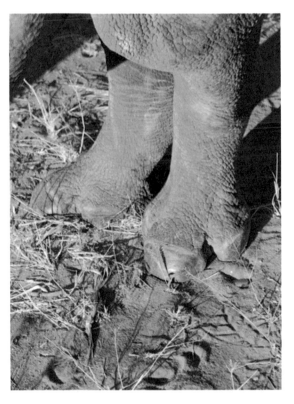

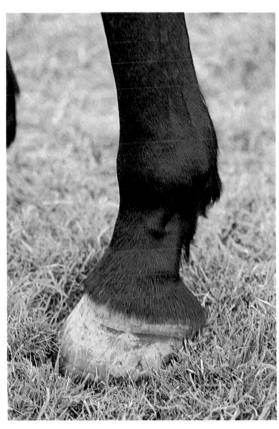

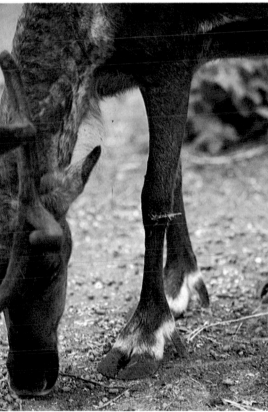

The hind feet of a rhinoceros illustrate one of the main types of hoofed feet. All mammalian feet evolved from an ancestor with five toes. In the most primitive condition, mammals stand on the whole length of the foot, but running is more efficient if the heel and instep are raised, so that only the lower surfaces of the digits touch the ground – this reduces friction and drag. The hoofed animals stand right on the tips of their toes, which have greatly developed nails, forming hooves.

(*Left*) Rhinos have tridactyl feet: the middle toe is greatly enlarged and carries most of the weight, while the first and fifth digits are so reduced that they no longer touch the ground.

(*Lower left*) The horse is even more specialized for running, and has a monodactyl foot – only the middle toe touches the ground and forms the entire hoof, giving even greater flexibility of movement. The large joint just above the hoof is the ankle joint.

(*Lower right*) The reindeer's hoof is made up of two main digits, the third and fourth being equally developed to give a didactyl foot. The second and fifth digits are still obvious, but no longer reach the ground.

feet into flippers (as in the seals) or lost one pair of limbs completely (as in the whales and dolphins).

Another feature which is common to most terrestrial vertebrate feet is that the five digits are tipped with claws. Like beaks and scales, claws are formed of keratin, derived from the basic skin layers. The flat nails on the fingers of primates like ourselves, and the hooves of many mammals, are modified claws. The cloven hooves of cattle, for instance, are the modified claws of the two digits that they walk on; and the hooves of horses have developed from the claw on their single remaining digit. Claws and hooves, like skin, are alive and full of blood vessels and nerves in their lower layers, but as they grow outwards the cells toughen and die to form a horny and insensitive layer. Often one side of a claw grows faster than the other, resulting in a curve. Claws, nails and hooves are designed to be worn away, and grow continuously. Just as the shape of the foot varies in different animals, so too does the shape of the nails, claws and hooves.

Every design of foot in the animal world has evolved for a particular purpose, to equip the animal to live in a particular place or move in a certain way. With the benefit of technology man makes shoes, effectively changing his feet to adapt to different kinds of terrain. Millions of years of evolutionary trial and error, successive minute changes adapted or discarded, have resulted in different feet, each adapted for a particular habitat or terrain.

It would be a lengthy task to try to describe all kinds of feet and show how their design fits their function. In general, however, animals that live or move on a certain type of terrain have feet with similar characteristics, and some of these general principles are outlined here.

Animals that specialize in fast running tend to have small, narrow feet. The bones between ankle and toe are elongated to make a more mechanically efficient spring lever, which moves in only one plane – forward/backward – so as to reduce the amount of muscle needed, and hence reduce weight. The running animal moves on its toes which are flexible and springy. The cheetah is one of the fastest animals in creation: compare its feet to those of the more sluggish lion. The cheetah's feet and gait are much finer and more slender than the broad pads of the lion, but they are not so versatile. The cheetah's feet look more like those of a dog than

The paws of a lion resemble those of most of the cat family. Cats and dogs walk in what is called the digitigrade position: the heel and instep are raised off the ground, making locomotion quieter and more versatile. The large pads on the ball of the foot and on the toes provide a cushion when walking and also help silence the feet. The lion has retractile claws – it can retract them while at rest or when walking, so that they do not catch in the ground and reduce his speed.

A pocket gopher tunnels through the soil in search of roots and bulbs to eat. Note the huge front paws and powerful fingers ending in strong curved claws for digging. A small animal, it can excavate burrows hundreds of metres long.

This red kangaroo is using its powerful tail as an extra support while standing still, like a living tripod. It has enormous feet, up to 40 cm long, which it uses as springboards as it bounds along at up to 24 km an hour.

a cat, and indeed this is the only member of the cat family whose claws cannot be retracted into folds of skin when they are not in use. The toes of runners are brought close together and reduction of the number of bones results in a stronger leg for the same weight. So retention of several toes must be favoured by other considerations, such as ability to grasp or to extend over an area. The ostrich is a bird that depends on running to the extent that it has lost the power of flight; it has only two toes, both long, narrow and flexible.

Jumping animals, such as the kangaroo, or the mouse-like desert jerboa, have enormous hind legs. The limb bones are lengthened to increase the spring mechanism, and when at rest the animal sits, as it were, on its whole 'forearm' from elbow to digits. The apparent large size of this rear 'foot' is exaggerated by the smallness of the front feet, which are not often used for locomotion at all. The foot of a great red kangaroo, which is about 1.5 m high, is about 40 to 45 cm long. A domestic cat has well-developed hind limbs for springing: it frequently sits on its lower hind legs, and the fur on the back of this section of leg is thick and rough to cope with the extra contact it makes with the earth. It is not such a marked adaptation as in the kangaroo,

The European mole spends most of its life burrowing through the ground to hunt worms and other small invertebrates. Its short stocky limbs bear massive feet which are broad and spade-shaped. These are used alternately to shovel back earth, using their powerful claws to break the soil surface.

for the cat walks and runs more than it jumps, but the similarities have a common cause. The *heel* of a horse is the hock – about 60 cm off the ground – and its *wrist* is the so-called knee of the fore-leg.

Quite different in character are the feet of the diggers, animals that habitually burrow into the earth. The friction drag of moving through the ground is potentially enormous, so the size of the limbs and the area through which they move must be kept to an absolute minimum; but at the same time, great strength is needed. The limbs of animals that lead an almost completely subterranean life, like the mole, are short and thick, and their feet are broad and powerful. Each short stroke of a foot must move as much earth as possible, and the mole's feet are spade-like with widely spaced digits.

In addition, the claws of digging animals are usually large, sharp and strong, to do the work of a pickaxe in breaking the soil surface. The aardvark of South Africa (its Afrikaans name, 'earth-pig', refers to its rather pig-like head) is a curious animal that digs for food in termite's nests. Its feet are short and massive with large, almost hoof-like claws on each toe. It is said that one aardvark can dig a hole faster than six men with shovels. Not only does it dig into termite nests to eat the insects, the aardvark digs burrows 4 m or more in length in which to hide during the day.

The armadillos of Central and South America are also powerful diggers, able to conceal themselves at amazing speed; they too have short, strong legs with daunting claws. The feet of the giant anteater,

another excavator of ant and termite nests, are not as massive as those of the aardvark. They are long and curved – so much so that the anteater is forced to walk on the sides of its feet with an ungainly bow-legged gait. The anteater is a scratch-digger, not a maker of burrows, so its claws do not need to be as large.

Diggers are not the only animals to have large claws. Often this feature denotes a predator that uses its feet to help trap its prey: the cats are a familiar example among mammals, and among the birds, the birds of prey – eagles, hawks, owls – have large hooked claws, talons. The sloth has big deeply curved claws too and yet no animal could look or behave less like a predator. Instead, the sloth uses its claws as hooks to help it hang upside-down from branches in the South American jungle.

Swimming animals often have webbed feet specially tailored to their chosen method of locomotion, though the bone structures vary considerably. In whales the finger bones are increased in number, and in ichthyosaurs, the whole 'hand' is reduced to a flexible plate of tiny bones. The radius and ulna of such swimmers – and this includes penguins – tend to be short and fat. Perhaps the most familiar example is the duck. The skin on a duck's foot, though quite leathery, is flexible enough to be folded when the foot is brought forward through the water, causing minimal drag: but it is strong enough to be stretched taut and pushed against the water on the backward stroke. The duck has four toes, arranged like those of most other birds into three forward-pointing toes and one pointing backwards: the forward three toes are joined by a web of skin, but

(Far left) The long elegant legs of the flamingo end in colourful webbed feet. The flamingo needs webbed feet for swimming, but it also needs them to prevent it sinking into the soft mud in which it often paddles while feeding.

The webbed feet of a duck, showing how the webbing stretches across the three forward-pointing toes, leaving the fourth, backward-pointing, toe free. The position of the legs well back on the body is also an adaptation to swimming – the duck is propelled from behind – but leads to a rather ungainly waddle when on land.

Caught in mid-air by the camera, this leaping frog has webbed feet which it uses for swimming. Like the duck, the frog is pushed forward from behind: compare the powerful back legs and their webbed feet with the slender front legs and webless front feet. When landing after a leap through the air, the webbing may also serve as a parachute to slow the descent.

The duck-billed platypus has a streamlined body and webbed feet front and back for swimming. On the front feet, the webbing extends even beyond the claws to form a fan-shaped paddle, but it can be folded back to allow the platypus to walk on land. The thick curved claws are used for digging its burrow in the river bank.

(a) Front foot of the platypus showing the webbed flap of skin in the swimming position.

the back toe is free so they can perch. Most other water birds share this arrangement, but in the pelicans all four toes are webbed.

Webs of skin between the toes are also characteristic of amphibians, especially the frogs, and to a lesser extent the toads; toads spend less time in water than do frogs. It was the long, webbed hind feet of the frog that gave man the idea of making rubber flippers to swim with, and hence coined the name 'frogman' for a diver. Among mammals, the otter's toes are webbed and the claws are short, contributing greatly to the lithe grace of its underwater movement.

A strange creature that seems to have the best of several worlds, where feet are concerned, is the duck-billed platypus. A virtual 'missing link' between the reptiles and the mammals, the platypus has features of both: it lays eggs, but nourishes its young with milk. It lives in streams in eastern Australia and Tasmania. Both sexes make burrows in the riverbanks for temporary shelter. The female also digs a long tunnel to a nest chamber.

The platypus has three-way feet. The five toes are long and slender, and can be bent into a 'fist' so that the animal can move – albeit without grace or speed – on land. The toes bear long, sharp, curved claws, typical of a scratch-digger, with which the platypus excavates its burrow. The rear toes are webbed like those of a duck; but the front feet are unique. The web extends along the sides of the feet and under, and beyond, the claws on all the toes *(diagram a)*. This more than doubles the surface area, and thus the paddling power, of the foot. One last peculiarity of the platypus is the sharp horny spur on the back feet of the males, through which poison from a poison gland inside the leg can be injected with a slashing kick. The spur is used in defence and against other males in contests, and the poison is strong enough to be dangerous to humans – not that many of them come into contact with platypuses.

Special problems confront the climbing animals which spend much of their lives in trees, or on rocks. Their feet must adapt to cling securely to a surface that may be small, irregular, unstable, slippery-smooth, or even (as in the case of a wind-whipped tree) in vigorous motion. The feet of climbers, therefore, tend to have long, wide-splayed digits that can cling to a flat surface with the help of hooked claws, or grasp around a branch. The feet of climbers are free of hair or feathers: naked skin with plenty of nerve endings is needed to adjust the grasp to the varied nature of the surface. Bumps or ridges

(Far right) The tree-creeper is able to run up or down a tree with equal ease. Two toes on each foot point forwards for gripping the bark on the way up, and the other two point backwards for gripping while going down. The long sharp claws hook into the bark, and its stiff tail feathers are used as a prop as it moves up the tree.

A chameleon moves slowly along a branch, one foot in front of the other. Its fingers and toes are fused into two groups on each foot, which grip the branch like a pair of pliers. When at rest, it will often curl its tail tightly around the branch for extra security.

on the skin also help the animal to maintain its position by creating a greater amount of friction against the surface of the tree, or other irregularities.

The Passeriformes or 'perching birds' have the typical bird foot: three toes forward and one behind, with which a bird can perch crosswise on a branch. A bird's sole is covered with rough bumpy skin, so that it can obtain purchase even on a small, weak, mobile twig which may be wet and slippery after rain. A few birds, such as the woodpecker, tree-creeper, and parrot, do not limit themselves to perching, but climb up tree trunks. All of these have two toes pointing forwards and two backwards. This means that they can climb not only up a sloping tree trunk, but down it as well, 'hanging' from the two rear-pointing toes instead of just one which would be too weak to support the body. Parrots also use their beaks as a third limb for gripping, and woodpeckers use their tails, which are large, and equipped with long, strong feathers on which to prop themselves. The tree-creeper is aptly named, being so agile running up, down or sideways on a tree-trunk that it may be mistaken for a mouse at a distance.

The squirrel is particularly well adapted for tree climbing. It has sharp claws, and instead of having

backward-pointing toes like the climbing birds, it can swivel the whole back foot round at the ankle so that it points backwards. The squirrel can thus hang from an almost vertical surface provided there is enough irregularity on the tree trunk into which to hook its claws. Many lizards do the same thing, turning the hind feet backwards and clinging upside-down with almost magical ease to a rock or wall that appears quite smooth.

Grasping hands for tree climbing are best developed among the primates. The opposable thumb, coupled with sensitive fingertips, gives the monkeys the freedom of the treetops. The chameleon has also developed opposable fingers. It seems to have two fingers, but in fact three digits and two digits have fused together, and the resulting two groups of toes can curl around a branch in a tight grip. The chameleon moves parallel to the branch, unlike the birds which perch crosswise. The chameleon's prehensile tail acts as a fifth grasping limb, giving it an unshakeable hold, as anyone who has tried in vain to unseat a chameleon will know.

Perhaps the least favourable arboreal surface is a shiny, dangling leaf. Yet tree frogs in tropical jungles can cling to such leaves, apparently defying gravity. They achieve this by exuding sticky mucus from their toe-pads, which glues them lightly to the leaf surface. The sticky effect is increased by pressing their mucus-moist bellies against the leaf.

The tokay gecko is a large lizard up to 35 cm long from the Far East. It feeds on insects and mice, coming out to feed at night. The skin on its back is quite soft, and is covered in rows of horny tubercles, but its feet and undersurface are tougher, to withstand the friction of the surfaces over which it moves.

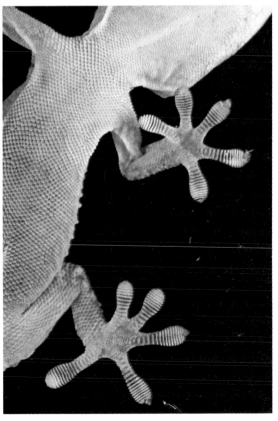

Perhaps the strangest feet of all belong to the geckoes, small lizards that may sometimes be seen inside houses in tropical regions. Not only can a gecko move quickly across walls and even ceilings, it can cling on to a vertical or even overhanging pane of glass. For an animal that is considerably larger than the flies it hunts for food, this ability seems virtually impossible. In the past it was thought that the gecko actually produced some form of glue on its feet, although that would have slowed its speed somewhat. The truth is even more extraordinary. Each of the gecko's toes has a pad bearing ridge-like scales. Under a microscope it can be seen that each scale bears hundreds of tiny, hair-like protrusions called setae. As if this were not enough, a further magnification shows that the individual setae are tipped by 'brushes' of up to 2000 incredibly small branched filaments, bearing saucer-shaped tips. This provides a phenomenal total of about 100 million points of contact. These minute saucer-tipped brushes can interlock with the tiniest irregularities that occur on even the smoothest of surfaces. The gecko holds on to the glass by surface tension alone. The attachments may be so strong that an attempt to pull a gecko from the glass will result in the glass breaking before the gecko lets go.

(*Far left*) Geckos are amazing climbers. Not only can they walk up vertical surfaces, but they can even walk upside-down across the ceiling. As this picture of the underside of a gecko on a window pane shows, the toes bear rows of supple ridges which act as adhesive pads.

It is important for a gecko to
able to cling tightly to the
surface on which it is stand
as it will often hunt and cat
quite large prey. Its toes ha
large surface area in contac
with the substratum, and th
ridges on their undersurface
can get a grip on minute
irregularities in the surface.
Geckos can even walk up
vertical panes of glass whic
appear far too smooth to ge
foothold on. This versatility
enables them to exploit eve
nook and cranny of their
habitat.

Fins and flippers

Fins are unique to fish, providing them with a very adaptable and efficient method of getting about in water. Fins are so useful that marine mammals (those which have evolved to live in water, instead of on land like most mammals) have developed the nearest thing to fins that they can: flippers.

All fish have the same basic complement of fins. But fish have evolved into a great variety of shapes, from the streamlined shark to the boxlike trunkfish, from the sinuous eel to the flattened plaice. It is not always apparent which fin is which; and certain species have dispensed with some fins altogether. A typical fish is shown in *diagram a*. There are three median fins, so called because they are set along the mid-line of the fish. These comprise: the dorsal fin on the back, which may be divided into two or more sections; the caudal or tail fin; and the anal fin, set behind the fish's anus. In addition to these there are two pairs of fins, the pectorals at the front of the fish's belly and the pelvics further back. (It was from these paired fins, or rather from the fin rays that support them, that the four limbs of the amphibians and higher animals evolved.) The fin rays are made of tough, plastic-like cartilage and are not part of the skeleton of bony fishes; they are attached to the

the muscles. These are ossified in the larger, faster fish such as Tuna.

A fish can move by using its fins alone, but not very far or fast. By paddling with its pectoral fins and waving its dorsal and caudal fins it will move gently forwards. For anything more positive, however, the fish flexes its body from side to side in an S-shape, thus pushing its caudal fin hard against the water to create a much more powerful thrust. At the same time, the dorsal fin acts rather like the keel of a boat, helping to maintain the fish in an upright position. By flexing the pectoral and caudal fins the fish can turn up, down, or sideways. The pectoral fins are also used as brakes, being pushed forwards like the flaps on an aircraft wing. The positions of the paired fins, especially the pelvic fins, are constantly adjusted to keep the fish from pitching or rolling; the pectorals tend to produce lift, which is counteracted by the downward thrust of the pelvics.

Fins are mainly used for movement, but they also serve other purposes. In some fish, such as the Siamese fighting fish or the guppy, the fins are enlarged and brilliantly coloured for sexual display the males showing off their colourful finnage

(*a*) Fins of a typical fish.

se look at a stickleback tail lows the radiating rays of age which support it – by g it contributes at least 50 ent of the thrust during ming. It also acts as a r. The black star-shaped les are pigment cells which xpand to change the colour fish.

These male Siamese fighting fish in full breeding colours are tearing each other's fins to shreds as they fight for supremacy in the competition for mates. The brilliant colours and large fins are designed to attract females, but simultaneously they antagonize other males.

The strange appearance of the lionfish *(below left)* is caused by its highly divided dorsal and pectoral fins. At close range its striking colours are a warning to would-be predators that it is poisonous: both these groups of fins can inject poison. The striped pattern also serves to break up the outline of the fish when viewed from a distance, a form of camouflage. The zebra firefish *(below right)* has a similar pattern, and its dorsal spines also contain venom. The membranes of its pectoral fins extend almost to the tips, giving the appearance of a pair of wings.

This gurnard illustrates the main fins found on fish. It has a pair of blue-edged fan-like pectoral fins, a brightly coloured spinous dorsal fin and behind it the soft dorsal fin, pelvic fins almost hidden by the pectorals, a whitish anal fin, and a caudal fin at the back. If its pectoral and bright dorsal fins were not erected, it would be well camouflaged on the sea bed. Such eye-like markings as that on the dorsal fin are often designed to distract a predator's attention from the fish's real eye, which is much more vulnerable.

fin rays into defensive spines, sometimes equipped with venom glands to make a vicious weapon. The dramatic lionfish of tropical seas positively bristles with poison spines, and as a warning of its dangerous nature is decorated with colourful stripes. Its battery of barbs means it cannot move fast, but then it rarely needs to make an escape: sensible sea creatures avoid it.

There is not enough space available here to do more than outline a few of the ways in which the basic fins have been adapted by different fishes. The shape of the caudal fin, for example, gives some indication of how fast a fish can swim. A fish with a large, squared-off caudal fin has more surface area to push against the water, producing greater thrust but also more drag; so such fishes tend to be generally slow-swimming apart from occasional 'sprints' – the pike is an example. The fast marathon swimmers of the ocean, like the tunny or the swordfish, have very stiff crescent-shaped tails. This shape also makes it easier for the fish to change direction suddenly. The tunny is the speediest of the ocean fishes, able to attain 70 kmph.

Pectoral fins show an enormous range of adaptations in different fishes. For a start, the extraordinary 'wings' of the skate and rays are

modified pectoral fins. The ray swims by undulating these pectoral fins with a wave-like motion. The body of the fish is flattened (though not in the same way as that of flatfishes) and the tail extends into a tapering cone or whip shape, which may carry a poison dart. There are some 340 species of rays, including the electric rays which can deliver shocks of up to 220 volts, the common skate, with its diamond-shaped fin profile, and the incredible manta ray, or devil fish. It is easy to see

The manta is the largest living ray, up to 6.7 metres across and 1360 kg in weight. Surprisingly, it feeds on small crustaceans and plankton, trapped on its gill rakers. Unlike most rays, its mouth extends across the front of its body, and the large mobile pale-coloured cephalic fins on either side of the mouth can be extended vertically. It has been suggested that these may form a scoop or funnel leading to the mouth while feeding.

A spotted eagle ray *(right)* and a cowtail ray *(below)* swim over the sea bed in search of food. Rays have large pectoral fins which extend to the front of the head and are used for swimming. Rays are relatively primitive fish, and have no swim bladder, which makes them prone to sinking unless they keep moving. The long whip-like tail acts as a rudder and is armed with venomous spines near the base for use in defence.

The English skate, seen here from the underside, is a scavenger, feeding along the sea bottom. For this it is well adapted, with its mouth on the underside. It swims by means of wave-like movements of its expanded pectoral fins. This is a young specimen, but when fully grown it may reach a width of 2.4 metres. It lives in quite deep waters, 30–600 metres, and feeds on fish, crustaceans and molluscs. The eyes, though visible from the underside in this picture, are actually on the top of its head.

(*Opposite, top*) A dead flying fish lies on the deck of a ship. They may glide as far as 400 metres using their enlarged pectoral fins.

Which fin is which? In fact, the plaice is lying on its side. As these fish develop, the bones of the skull twist so that both eyes are on the upper surface. They swim by undulating their fins.

pectoral fins are enlarged and elongated to form gliding 'wings'. Flying fish are quite common in tropical oceans and are often seen from ships; in fact they have been known to misjudge their 'flight path' and land on deck by mistake. The flying fish takes to the air usually to avoid the pursuit of predatory fish. The pectoral fins do not usually flap to provide thrust; the fish takes off by vigorous tail movements and bursts through the water surface, or it 'taxis' on the surface by flicking the elongated lower lobe of its caudal fin in the water. Once airborne, it may glide as far as 400 m before falling back into the water. If caught by air currents, the fish may reach a height of 5 m or more during its flight.

The strange little fish called mudskippers have modified their pectoral fins to emulate not birds, but land creatures. The pectoral fins are borne on short 'arms' and may be turned sideways and backwards, as well as forwards. The fin rays are stiff and can support the front of the fish rather like a man doing press-ups on his fingertips. In this way the mudskipper can hop on to the mud at the edge of the shallow water it inhabits. It seems probable that the first amphibious animals emerged from the water in a similar way, increasing the supportive strength of their pectoral and pelvic fins until they gradually evolved into limbs.

The flatfishes look rather like the rays, but the fins that fringe their bodies are adapted not from the pectoral, but from the dorsal and anal fins. Instead of being flattened vertically, like the ray, a flatfish such as the plaice is flattened laterally. When it rests on the sea bed it is lying on one side. The dorsal fin runs the length of the body from above the eyes to the tail, and the anal fin nearly as far on the other side. The strange thing is that in their immature stages, flatfishes are like other larval fishes; but at a certain point, one eye moves round the head to lie beside the other, the mouth also twists round to that side, and the fish turns sideways on to the world. Usually only the upper surface is coloured, the underneath or 'blind' side being white. Apart from undulating their fins to swim, most flatfishes can use them to stir up the sand or pebbles on the sea bed and partially cover themselves to camouflage their outline.

One cunning adaptation of the dorsal fin can be seen in the angler fish. This squat fish is camouflaged to merge into the background of the sea bed, where it lies in wait for smaller fish prey. The first ray of the angler's dorsal fin is set independently forward on the fish's head like a miniature fishing

why the manta incurred its other name, for its triangular pectoral fins are jet black and can reach a total width of 6 m. In the tropical waters where it lives, the manta ray may leap 2 m clear of the water and sail through the air like a giant bat, hitting the water again with a loud boom. Despite its terrifying appearance the manta ray is harmless to man, feeding on small fish and crustacea, although it is so strong that if accidentally hooked it can drag the fishing boat through the water.

A fish that is even better known for its habit of leaping out of the water is the flying fish, whose

Seen underwater a dolphin could easily be mistaken for a fish. One of the main features which distinguish marine mammals from large fish is the tail, which bears a pair of horizontal flukes: the caudal fin of fish is always vertical. Unlike fish, the dolphin uses powerful up and down strokes of its tail flukes for swimming; the flippers and dorsal fin are used mainly for steering.

rod, tipped with a small fleshy 'lure'. Other fish, drawn to the lure by greed or curiosity, do not notice until too late that it hangs right in front of the angler fish's capacious jaws ... Species of angler fish that live in the perpetual darkness of the ocean depths have luminous lures to attract their prey.

Perhaps the strangest dorsal fin of all, and hardly recognizable as such, is that of the remora. This fish appears to have a flat, oval, ridged plate on its head. The plate acts as a sucker by which the remora can attach itself very firmly to the underside of an object – sometimes a ship, but usually the belly of a whale, shark, or turtle. The sucker plate consists of the rays of the spiny dorsal fin, divided and flattened to form ridges, which may be raised or lowered to create suction. By hitching a lift on a shark the remora steals protection, free transport, and crumbs that fall from its host's table.

Many people may think that the dolphin is a fish, instead of a mammal. Compared with a shark, dolphins also have the elongated, smoothly tapering outline that we call streamlined; both have a roughly triangular fin on the back and a curved tail fin, and two broad flipper-fins on the lower front part of the body. The shark, however, has the piscine characteristic of breathing water through its gills, whereas the dolphin breathes air through lungs; and the tail fin of the shark, like that of all fishes, is set vertically, while the tail (or fluke, as it is known) of the dolphin is set horizontally. Also, if one dissects away the tough skin on both animals, it is evident that the shark's fins contain cartilaginous fin-rays; but the dolphin's dorsal fin and tail are formed of tough fibrous tissue, and its flippers contain bones making the five digits of typical mammals. The shark also bears other fins on its body – all supported internally by fin-rays – while the dolphin has none. The shark, whose prehistoric lineage stretches back millions of years, perfected its design for aquatic life and locomotion long

This blue angelfish, an inhabitant of Caribbean coral reefs, has a very characteristic body design. It is a tall thin fish, laterally flattened, which enables it to glide into crevices in the reef. Its much extended dorsal and anal fins give it great vertical stability, and the pectoral fins, just behind the head, can be used as brakes.

Flippers are not the best tools for scratching an itch, but they are all the sea lion has. Sea lions are aquatic mammals with streamlined bodies, but have retained their hairy coats and all four limbs, and can move about on land more easily than seals. As can be seen in this picture, their limbs (flippers) can still be flexed, and the separate digits are visible although they have lost their nails. The forelimbs are quite long, and unlike whales and manatees, sea lions have no flukes at the end of the tail, the hind limbs being used for swimming.

before the dolphin's ancestors evolved. Much later, the dolphin developed to a similar pattern when it returned to a totally marine lifestyle. Having no fins, it gradually lost its hind limbs and turned its forelimbs into flippers, growing the nearest thing to dorsal and caudal fins possible.

Among the marine mammals, the cetaceans – whales, dolphins and porpoises – are nearest to fishes in form. The other two main groups of marine mammals, the sirenians (sea cows, manatees, and dugongs) and the pinnipeds (seals and sealions) are also fishlike in outline. The sirenians have stiffened fore flippers, like those of the dolphin, and a paddle- or fluke-shaped tail; as in the cetaceans, the pelvic girdle and the hind limbs have been almost entirely lost. In fact, apart from their mammalian, seal-like heads and the lack of a dorsal fin, the sirenians are similar in form to dolphins. (The head of the manatee, which looks rather like a walrus without tusks, is hardly beautiful; and yet these were the creatures called mermaids by the early sailors, the enchanting 'sirens' that lured men to run their ships aground on rocky reefs.)

The name 'pinniped' or 'fin-footed' derives from the limbs of the seals and sealions. These mammals, although almost wholly aquatic, do return to land to breed, and they have therefore retained their limbs. When they are swimming, the hind flippers are laid back to look like a tail, and the fishy shape is very evident. Their feet contain the five bony digits of the typical mammal, encased in thick rubbery skin to give a flipper shape. The flippers, however, are not rigid as in the other marine mammals; and in most species there are claws projecting from the digits, quite unlike the dolphin's flippers. In fact the flippers of a sealion look rather like bony hands encased in floppy, rubber gloves. In the water, the fore flippers are used like oars for propulsion, while the hind flippers are used like a tail, to steer and thrust.

Apart from the fact that sealions (or eared seals) have external ears and the true seals do not, the two groups may be distinguished by their flippers. The sealion can twist its fore flippers backwards at the 'wrist', and bring its hind limbs under its body to sit on its haunches in a rather dog-like manner, or move in an ungainly but quite rapid gait. The true seal cannot do more than waggle its flippers, and on land it is limited to slow and ineffectual movement by rolling or lurching its body. It is, of course, always a sealion that is seen in a circus, not a seal: usually it is the smallest species, the Californian sealion. Circus performances may demonstrate the intelligence and agility of these creatures, but they cannot show their superb grace for doing what they were designed to do – swim.

The Florida manatees are so adapted to underwater life that they cannot come out on land. Manatees have lost most of their hair, and have only vestigial hind limbs and pelvis. The fore-limbs have become flippers, and broad tail flukes are used for swimming. Manatees live in shallow water near the coast, grazing on aquatic vegetation.

Wings

All sorts of animals apparently fly. Even fish, frogs, squirrels and snakes may be encountered in the air. The only animals that really fly, however, are those with wings. The others are all gliders, equipped with mechanisms with which they can descend, more or less gracefully, from a high point to a lower one. But none of them can fly upwards. Wings are vital for true, powered, flight.

Thousands of species, from no less than four major animal groups – insects, dinosaurs, birds, and mammals – have developed wings. The invertebrate insects were the first flyers, taking to the air at least 280 million years ago. Given this incredible span of time for evolution, one might have expected insects to improve their flight efficiency and potential to such a degree that they could have dominated the skies to the exclusion of all other creatures. Instead, the earliest known flying insects were very similar to the dragonflies that exist today.

The next animals, and the first vertebrates, to grow wings were the pterosaurs, the flying dinosaurs. Pterosaurs developed into many different species, and inhabited this planet from the Triassic period until the end of the Cretaceous, about 65 million years ago – a dynasty enduring some 70 million years. The largest pterosaur known from fossil evidence was *Pteranodon*, a terrifying creature with a huge 'beak', a large horn jutting from the back of its skull, and a wingspan of about 8 m. Other

A desert locust on a flight that may carry it thousands of kilometres. The tough forewings have camouflage colouring, and when at rest conceal the larger folded hindwings. The latter are much more flexible, and can be seen here twisting to help lift the insect over the twig. Its wings beat relatively slowly, only 12–15 beats per second.

196

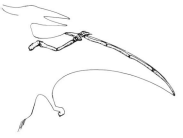

(a) Arrangement of wing bones in *Pteranodon*.

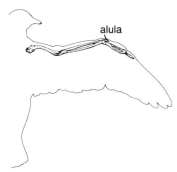

(b) Arrangement of bones in a typical bird wing.

A blue-tit coming in to land at its nest has reached stalling speed: the air turbulence over its wings is about to cause it to drop to the ground. It compensates by raising its alula or bastard wing, as in this picture. This forms a slot through which air rushes, restoring a smooth fast airstream and preventing stalling.

pterosaurs though not so large, were equally ugly and ferocious in appearance. As far as we know, the wings of pterosaurs were constructed rather like those of bats, with a tough elastic membrane stretched between body and arm bones. The arrangement of the bones, however, was quite different to bats *(diagram a)*. Pterosaur wings were supported by an enormously elongated 'little finger', and the animal had a clawed hand half-way along the leading wing edge. The main problems associated with such a wing might have been that a hole or tear in the membrane could have permanently grounded the pterosaur; and that once grounded, it was limited to a slow, shuffling, and vulnerable gait.

During the pterosaur reign, the first birds developed; in fact many scientists now believe that birds *are* warmblooded dinosaurs that have survived to modern times. In the typical bird wing, feathers have taken the place of a skin membrane, and the bones of the 'hand' *(diagram b)* are reduced to two fused digits. Approximately where we have a thumb, a bird has an alula, a tiny bone supporting a tuft of short feathers at the front wing joint. The alula actually plays an important part in flight by correcting airflow over the wings to help prevent stalling at low speeds. The strength, lightness, and flexibility of feathers enables birds to fly with more power and versatility than bats or insects, making them truly lords of the air.

197

An expert in soaring thermals, the turkey vulture has deeply slotted wing-tips which minimize turbulence at slow air speeds. The thermals enable it to soar to heights of several thousand metres, where it can scan vast areas for carrion. By gliding from one thermal to the next, it can travel hundreds of kilometres with relatively little effort.

More recently – about 50 millions years ago – some small shrew-like mammals developed wings, and became the first bats and the only flying mammals. Bat wings consist of a double layer of skin, rich in blood vessels, stretched between body and arms and also, in some species, extending to the tail. Unlike other vertebrate wings, those of the bat are supported by the splayed and much elongated 'fingers' of the hand *(diagram c)*. Even the thumb remains as a hook with which the bat can cling to a surface. Apart from fast and agile flying, a bat's wings are vital in maintaining the correct body temperature: they act as radiators, to get rid of excess body heat produced by flight. While birds and insects have legs to walk on land as well as wings to fly, bats are adapted for flight alone, and make awkward landlubbers.

In order to achieve flight, any wing must fulfil certain complex aerodynamic principles, which can only be outlined here. Basically, if the air pressure on one side of an object is lower than on the other, the object will tend to move towards the low pressure area. Air flows over a convex surface faster than over a flat one, creating an area of lower air pressure. So, if the upper surface of a wing is curved upwards, the flow of air over it will cause it to lift. This is the essence of the aerofoil shape used

(c) Typical arrangement of wing bones in bats.

(Far left) This Nigerian fruit bat and its baby are mainly nocturnal, and have large eyes for seeing in the dark. When at rest, their folded wings form built-in sleeping bags. As well as the main thumb hook *(see diagram c)*, most fruit bats have another claw on the second digit, clearly visible here.

A violet-cheeked hummingbird sips nectar from a passionflower. Hummingbirds can fly upwards, downwards, sideways and even backwards, but the rapid wingbeats needed to hover in front of a flower call for a lot of energy. This is provided by the sugar-rich nectar, a very concentrated source of energy.

in the wings of airplanes as well as animals. Tilting the leading edge of the wing upwards into the wind increases the lift, but at a certain angle – the stalling angle – the airflow becomes irregular or turbulent, and the lift is lost. The effect of lift is counteracted by that of weight, so any flyer must be as light as possible; and the forward momentum of thrust – produced by the wings pushing against the air – is counteracted by the backward pull of drag. The

Like the turkey vulture, gannets make use of updraughts of air, in this case air currents around a cliff face. Once they have been carried to a good height, they glide on the air currents looking for fish in the sea below. Their long narrow wings and streamlined bodies are adaptations for high-speed gliding.

efficient flyer, therefore, must achieve the optimum balance between these forces to suit its environment and lifestyle. For example, long narrow wings, such as those of the albatross, fly fastest with least effort, but tend to be very heavy and allow little manoeuvrability. Broad wings, like those of the vulture, provide plenty of lift – the effect is of wing tip vortices, due to the difference in pressure above and below the wing. With a short wing, a lot of lift is lost. With a slotted wing tip, the vortex can be converted into thrust – and this has been demonstrated on airplanes. So vultures' wingtip feathers (the primaries) can be separated like venetian blinds to allow the air to flow beween them. Also the vulture in flight flaps its wings as little as possible, preferring to ride the skies on thermals – rising currents of warm air. By contrast, the short wings of a hummingbird make a figure-of-eight wingbeat as much as 50 times per second, enabling this aerobatic virtuoso to hover, take off or land vertically, and even fly backwards.

A great spotted woodpecker takes off from its nest hole. The tail feathers and wing feathers are spread wide, and the alula or bastard wing can be clearly seen as the wings prepare for the first downbeat. Although a bird can push off with its feet when taking off, it needs to beat its wings hard to acquire sufficient lift to remain airborne. The woodpecker's nest hole is well up in the oak tree, so it can afford to glide downwards a little as it starts to fly.

201

Insects, the most numerous flyers, are not the most efficient in general; few species can fly faster than 18 kmph. They usually have a cumbersome four wings compared with a bird's two, although many four-winged insects use their wings as a pair, with a kind of zip-fastener mechanism to link the wing together. Unlike the wings of birds or bats, those of insects are not directly moved by muscles but by changes in the shape of the thorax (the centre section of the insect's body). As the thorax contracts, the wings lift; as it expands, the wings flap down. Flight takes a great deal of energy. An hour's flight may cost a fly up to 35 per cent of body weight. Locusts can fly for 8 hours at about 10 kmph and then replenish the lost fat in another 10 hours. Few birds can fly more than 48 hours non-stop. Bees, wasps and flies can beat their wings more than 200 times per second; but no flying insects can compare with the midge. This tiny troublesome creature may beat its wings as much as 1000 times per second, causing that ominous whine as it comes for you at 2 kmph.

The original master of the vertical take-off – a green lacewing prepares to loop the loop. The secret of its skilful manoeuvring is that the two pairs of wings can be controlled independently of each other, and can even be clapped together either above or below the body.

The common sympetrum dragonfly *(right)* and the green lestes damselfly *(below)* are descendants of a very primitive group of flying insects. They are skilful flyers, catching their prey, usually a mosquito, on the wing. The two pairs of wings beat almost exactly out of phase. This produces a smooth flight, but causes considerable air turbulence, which reduces efficiency. Like their fossil ancestors, these insects are unable to fold their wings when at rest.

Horns and antlers

Thousands of rhinos are illegally killed each year to serve the market in rhino horn. Solid horn is used for ceremonial dagger handles in the Yemen, powdered horn is used as a fever cure in the Far East, and as an aphrodisiac in some parts of Africa. The very adaptation designed to defend the rhino against its enemies is now the cause of its destruction.

The antlers of this wapiti, or American elk, are still in velvet. Elk shed their antlers in the autumn, and grow them anew in spring. While they grow, the velvet skin nourishes them. A wapiti's age is revealed by the size and complexity of his antlers – each spring he grows a larger and more branched set than that of the previous year.

A sorry-looking sight, a caribou stag has almost shed his velvet. The soft tissue that nourished his growing antlers is no longer required, and the antlers will soon lose their reddish stain and be ready for the rutting season. It is the drying up of the blood vessels underlying the velvet which causes it to peel, and their former channels can be seen as grooves in the completed antlers. The peeling velvet seems to irritate the stags, and they are often seen rubbing their antlers against trees to get it off.

developed for the mating, or 'rutting', season. This indicates that the main reason for their existence is connected with sex and not, as one might think, for self-defence. Defence is obviously a valuable use for antlers, as can be seen when a reindeer confronts a wolf, but if defence were the most important factor one would expect both males and females to possess them. During the rutting season there is intense rivalry among the stags for the privilege of mating with groups of females. In combat, the antlers are often used as grappling devices by which two stags can lock together and push or twist until the superior strength of one of them is apparent. Occasionally two stags become inextricably locked together and eventually die. As the social hierarchy of the deer herd becomes defined by such sparring, a dominant stag often need only display his larger antlers – perhaps with a toss of his head – for rivals to concede. Without a fight he is acknowledged the master.

Many of the prehistoric ancestors of deer had antlers – some, the protoceratids, having antlers on their muzzles as well as their heads, and tusks, too. But the most impressive deer of all was the giant deer. As the best remains of them were found in

Irish peat bogs it is often called the Irish elk. This deer was larger than any alive today, and had massive antlers spanning 3 m or more and weighing over 30 kg. It finally became extinct about 500 BC (though much earlier in Ireland). There has always been controversy about the reason for their extinction and for the size of their antlers. Opponents of Charles Darwin's theory of evolution, (which holds that adaptations must be beneficial to the individual in order to be perpetuated), suggested that the giant stag's antlers developed to the point where the animal could no longer hold its head up. This theory was countered by that of allometry (proposed by Julian Huxley) which points out that the growth rate of antlers increases relative to the growth rate of the deer; so the giant deer, simply by being big, had proportionally larger antlers than smaller species of deer. The antlers of the giant deer present maximum surface for frontal display, implying that the stags with the largest antlers became dominant and more successful at reproducing, and therefore reinforced the trend for large antlers. Growing these huge antlers every year, however, must have placed an extraordinary strain on the animals.

A bull moose in the rutting season, resplendent in his newly grown antlers. On a mature bull, the antlers may have as many as 40 points and measure almost two metres from tip to tip. Moose bulls, who lead solitary lives for most of the year, will fight each other for possession of the females during the rutting season, locking antlers in a contest of strength. Usually the injuries inflicted are not fatal, but for an old male the strain of many rutting season fights may weaken him and reduce his chances of surviving the winter to come.

Tails

The tail on an animal often looks like an after-thought, an apparently useless remnant that was forgotten while other more important and obvious parts of its design were being perfected. It is certainly true that many animals with tails seem to be able to survive, even thrive, after losing their tails accidentally. Many animals that have no use for a tail have lost them during the course of evolution: we human beings are a prime example. And it is quite remarkable how many different functions different tails may serve.

There is the simple, familiar fly-swatter tail, as seen on horses and donkeys, cattle, and larger animals including elephants. Sometimes it seems that the tail is not long enough to reach all the parts where flies may settle, and horses solve this problem by standing close together, nose-to-tail, so that a flick of one tail serves two animals. (A horse also uses its tail as a balance and as an aid in running and jumping.) The tail of the elephant, though a metre long and tipped with tough bristles, is rather insignificant in relation to the size of the beast, and could not reach all the flies that settle on it: but the long trunk at the front end helps.

Some animals have tails that can kill more than flies. Their tails are weapons, able to threaten creatures up to their own size or even larger. We can be thankful that the glyptodont *Doedicurus* no longer roams the earth, for this 4-metre-long prehistoric creature, resembling a giant armadillo, had a tail like the mace of a medieval knight, tipped with thick, sharp spikes. One stroke of the glyptodont's tail would have caused terrible injuries. A modern equivalent, though on a smaller scale, is the porcupine. Its tail, like its body, is covered with long spines. The porcupine is likely to shake its tail in the face of anything threatening, such as a hunting dog or a leopard, leaving its assailant with a lip full of barbed spines that work into the flesh and create festering sores. Some animal tails are powerful enough to constitute a formidable weapon, even without spikes or spines: such is the tail of the alligator. A favourite predatory ploy of the alligator is to catch its prey, such as a deer, by the nose while it is drinking at the riverside, and then to pull it into the water and thump it with its broad, muscle-packed tail to wind or stun the deer as it struggles. Alligator hunters know the strength of their quarry's tail as well as the number of its teeth, and keep an eye on both ends of the creature. Another tail weapon belongs to the thresher shark, which has a scythe-shaped tail as long as its body.

A mare grazes with her foal among the early spring buttercups. Horses use their tails as fly swats, flicking away irritating insects. This foal has not yet grown a very long tail, so its mother is sweeping her long tail over its back.

An American alligator basks in shallow water, its mouth slightly open to keep cool, and its powerful tail resting on the mud. While swimming, the flexible tail is an important source of thrust and also acts as a rudder. But it is also a lethal weapon, used to stun its prey. It can even be used to flick fish towards its waiting jaws.

The European spotted salamander is probably very similar in shape to some of the earliest amphibians, with a long muscular tail and short legs. The male salamander's tail has one rather novel use, however – to act as a noose to hold his mate in the correct position during fertilisation.

Sea horses have prehensile tails. They can be coiled around corals or vegetation to stop the fish drifting away. The sea horse actually swims in an upright position, using its dorsal and pectoral fins, but as can be imagined, this is not very efficient.

(Far right) Spider monkeys can use their tails as an extra leg for their aerial acrobatics. Only New World monkeys have this ability, which requires powerful tail muscles. The tail is also used to explore crevices for food.

Almost all the Australian possums are tree-dwellers, and have prehensile tails, but the scaly-tail possum has a tail that looks as if it has rubbed all the fur off. In fact, this is the way its tail grows – the scaly surface gives it a better grip on the branches.

With this it thrashes the water round shoals of prey fish to confuse and stun them.

Tails may be of special importance in movement. The most obvious locomotory function is that of the prehensile tail, used most effectively by monkeys but also by chameleons, opossums, some lizards and salamanders, pangolins and other animals. The sea-horse is the only fish with a prehensile tail, which it uses to cling to fronds of seaweed as it is not a strong swimmer. The most versatile prehensile tail is probably that of the spider monkey. Its tail is even longer than its thin spindly arms, and is used with just as much facility as the limbs. The spider monkey can hang by its tail alone, swinging to reach a tempting fruit in the jungle treetops; or it can hang from one hand and use its tail to reach, grasp, and bring back the fruit. The end section of the spider monkey's tail is covered with naked skin,

(*Overleaf*) A privet hawkmoth caterpillar prepares to attack a new leaf. It is typical of hawkmoth larvae – the light diagonal stripes which break up the body outline from a distance, the coloured spiracles, a slightly lighter colour on its back than on its belly, and a characteristic soft yellow and black spine or 'tail' at its rear end. Some hawkmoth larvae have a defensive display in which this threatening tail, although harmless, is used to discourage a would-be predator. Yellow and black coloration is often used as a warning of poisonous flesh in the insect world.

A stiff tail makes a useful 'third leg' for a great spotted woodpecker as it drills at a tree in search of grubs. When in this position it is otherwise anchored by only two of its four clawed toes, since the other two point backwards to help it grip when going down a tree. A good grip is essential if it is to be able to drill with any force.

A pair of kangaroo rats from the deserts in the southwest of the United States demonstrate two of the uses of their long tails. The one on the right is using its tail as a third leg while it stands up on its hind feet, resting the bushy tip of the tail on the ground. The other has just leapt on to the scene and its tail is still in the counterbalancing position.

ridged for better gripping, and is as sensitive and dexterous as the monkey's fingers.

Animals that habitually run on two legs instead of four often have a long, flexible tail for use as a counterbalance. The tail of a kangaroo acts like one arm of a cantilever counteracting the forward thrust of its body weight when it leaps. When the kangaroo is at rest it leans back on its tail as if it were a third leg. The basilisk lizard also uses its long tail as a counterbalance when it runs on its hind legs only. Long-legged desert mice, such as the jerboa, which jump like kangaroos, have long tails for this purpose. The jerboa can change its direction

– even turn right round – in mid-leap, simply by moving its tail. Even the tail of the domestic cat is important to its balance when it jumps.

Birds' tails, apart from helping to counterbalance their bodies when they are standing or perching, are important in flight for steering and braking, acting rather like the rudder of a sailboat or the landing flaps of an airplane. In water, too, tails are useful aids to animal locomotion: the otter sweeps its muscular tail as it swims, to provide both thrust and steering. A few birds, notably the woodpecker, use their strong tails as a strut or third leg while they cling to a steeply inclined tree trunk.

Other birds have tails which are designed less for flight than to be admired, and very eye-catching some of them are. The superb tail of the peacock is a familiar example. Many species and varieties of pheasant also have beautiful tails to complement their rich plumage. Some Oriental cockerels have been bred with trailing tail feathers several metres long, like shimmering ribbons. The Australian lyre-bird has a magnificent tail, lyre-shaped as its name suggests. In New Guinea, several of the birds-of-paradise have tails of unequalled splendour, fountains of feathers in brilliant and iridescent colours. Some birds-of-paradise increase the drama

Probably the best known of all courtship displays, the extravagant tail of the peacock can be quite an encumbrance when it is not courting, even when folded back in a train. There may be as many as 150 'eyes', each on a separate feather. These are not tail feathers, but upper tail covert feathers. The barbules of these feathers are unhooked so that the barbs stand by themselves.

Clinging somewhat uncertainly to a wheat stalk, a harvest mouse anxiously contemplates its next move. Like many mice, it uses its tail as an anchor when on slender stems. Harvest mice live in cornfields, and the wheat ears are one of their main foods in summer, so they are skilled little acrobats. When running down stems, the coiled tail is an effective brake.

You might think that the bright blue tail of this Sudanese skink would attract the attention of a predator. It is certainly the best end to attack from the skink's point of view. If the tail *is* seized, the skink can detach it and make a quick getaway before the predator realizes that there is no longer a skink on the end of it. Eventually the tail will regenerate, although it may not grow quite so long next time.

The beaver's tail is covered in scales, and is used as a rudder when the beaver is swimming, steering it left or right, up or down. It is also used to sound the alarm if danger threatens: the beaver slaps it down hard on the water surface to warn other beavers in the vicinity.

an escape. In some species the distracting effect is increased by the detached tail being able to wriggle like a worm. The tail may even be brightly coloured to be still more eye-catching. Lizards with disposable tails cannot only move and function quite well without them, they can usually regenerate the lost portion. Some mice are also endowed with tails that may be jettisoned during escape, but unfortunately they cannot grow replacements.

There is a wide range of animals which use their tails as social signals, particularly animals that live in groups. Ring-tailed lemurs carry their black-and-white striped tails stiffly erect when they are feeding on the ground; in the trees their tails may hang down conspicuously below the branches. Either way the tails provide instant visual information about the numbers, age and status of lemurs in the group at any given moment. Some deer have white tails which, when flicked, are easily seen at a distance, when the animal's body may be camouflaged. The beaver, another social animal, uses its tail as an auditory signalling device: it slaps its paddle-shaped tail on the water surface, making a loud clap to raise the alarm.

Members of the canine family make constant use

of their tail display by hanging upside-down from a branch, the better to flaunt their finery. All these birds display their tails in mating rituals, to impress females and snub other males.

The tails of some lizards are also designed to be seen, but for a different reason: deception. Their long, whip-like tails break off from the rest of the body very easily. The purpose of this is to distract a predator at least long enough for the lizard to make

Like striped scarves waving at a football match, the tails of these ringtailed lemurs are held high in the air when the animals are foraging on the ground. Since ringtailed lemurs live in groups, the tail may be used in communication, as a form of social signalling. Certainly it makes it easier for the lemurs to see each other in shadowy parts of the woodland and at dusk. The tail is very long to be trailed on the ground, yet it must take quite a lot of energy to keep it erect. When up in the trees, the lemurs do allow their tails to hang down.

A meerkat suns its tail. Meerkats (suricates) live in communal burrows, rather like prairie dogs, in the dry veld of South Africa. They can usually be seen at the burrow entrance, standing on their hind legs, using their tails as props, to scan the area for enemies such as hawks and eagles. They hunt in groups for insects and small vertebrates, purring to each other to keep in contact.

of their tails for social signalling, as is evident among domestic dogs. A dog holds his tail high and stiff to show dominance; low, even tucked under the body, to show a fearful or submissive demeanour. To indicate friendliness it wags its tail vigorously from side to side. Dogs which have had their tails amputated in the name of fashion, such as boxers or spaniels, must wag their whole hind quarters in a desperate attempt to prove their goodwill.

There are plenty of other, less common, functions of a tail, just a few of which may be outlined here. The fat-tailed sheep of the Middle East stores food reserves in its tail to draw on when food is scarce. Some lizards also store fat in their tails. The hippopotamus wags its tail during excretion, scattering its droppings over a wide area to leave a pungent marker on its territory. The caterpillars of most hawk moths have a slim, sharp-pointed 'tail' at the end of their bodies; although this tail is quite harmless, it looks like a large sting and may help to discourage predators such as birds. The puss moth caterpillar goes one better, having a pair of red whip-like tails which can be lashed back and forth over its body to drive off parasitic flies.

The tailless amphibians – frogs and toads – have no need of tails for terrestrial movement, but their larvae (tadpoles) live in water and swim with the aid of their oar-like tails. These tails are absorbed back into the body during the process of metamorphosis by which the tadpoles become frogs.

The krait is a very poisonous, yet unaggressive, snake: if threatened it raises its tail in a curve like a cobra about to strike. There is even a red patch on the tail which looks like a warning red 'throat'. Finally, the amphisbaena, a legless reptile rather like a slow-worm, has a tail which is blunt and rounded, very like its head. The South American natives call it a 'two-headed snake'. The effect is certainly confusing, for a predator catching sight of the amphisbaena cannot tell which end is which, or which way it is going.

A tail with a sting in its tip. The scorpion uses its venomous sting not only in self defence, but also to paralyse its prey, which it usually catches with its large pincers. The venom is a neurotoxin, produced by glands lying in the base of the sting. This species is quite poisonous, and its sting can be fatal to old people and children.

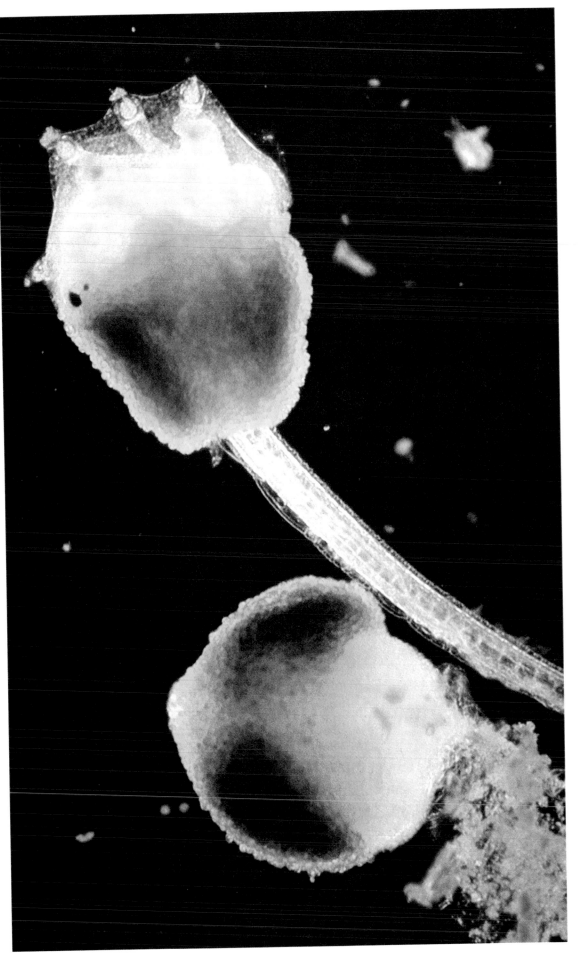

Tadpole-like larvae of sea squirts present several fascinating problems for the student of evolution. The vivid green colouring is due to the presence of a curious group of algae in the larvae's tissues. Until recently this group of algae was unknown to science, and to date it has been found only in sea squirts. It appears to be a symbiotic relationship, but the benefits to each partner are not fully understood. Young tadpoles actually swim towards light, perhaps to help their algal partners.

The ascidian tadpole's other claim to fame is its relationship to the vertebrates, and ultimately to man. It is one of the few invertebrates which have a segmented 'backbone' above which is a nerve cord running to the brain. We cannot know if the sea squirt tadpole is a direct link in the evolution of vertebrates – all we can say is it is one of the few living invertebrates with a segmented backbone.

Postscript

HENRY BENNET-CLARK

It is easy to think of animal design as simple. It is not. It may however be refined, ingenious, appropriate and elegant. These are qualities that can only be achieved as the result of continual change and improvement.

The design process has produced elegant systems in great variety but there are other things it has failed to do. These are just as instructive as the successes.

No animal can complete its life cycle without feeding. Vital processes consume chemical energy. Many animals, however, can exploit symbiotic living plants to provide most of their energy. These associations are fairly unusual and the animals that exploit them take on some of the characters of the plants; they require sunlight and must provide the plants with the salts, nitrogen and carbon dioxide that photosynthesis requires. Symbiosis of this type imposes severe limits on the life of both the animal and plants that are associated. As a result, this life style is not as adaptable as many others. Most animals seek food, be it plant or animal – this is a major factor in their design.

Though their movement seems efficient, all animals suffer from energy losses because their limbs swing forwards and backwards. Because starting and stopping a swinging limb uses energy, limbs have a natural inefficiency. The losses can be reduced by having springy tendons on hinges in the limbs to store and re-use energy but limbs are still less efficient than a moving wheel, where energy is only lost through friction in the bearings. Although bacteria did invent the wheel, or at least a rotating propellor which they use in swimming, animals have not yet succeded in doing so. The reason, presumably, is that solutions to the problems of getting blood vessels in and out of a water-tight, load-bearing, growing, rotating joint really are rather difficult to evolve!

Another limit of design is mortality. Although animals can repair certain types of wear or damage, they do not live for ever. Part of this is by design and part from failure to repair. For many animals, there is clearly an optimum life-span; in man, it obviously cannot be less than twelve years and one could envisage problems if everyone lived, like Methuselah, nine hundred and sixty-five years, at any rate with prevailing human patterns of society. The optimum is somewhere in between and subject to natural selection. Within a life span it may be advantageous to repair the defects, but immortality is the last object of the grand design.

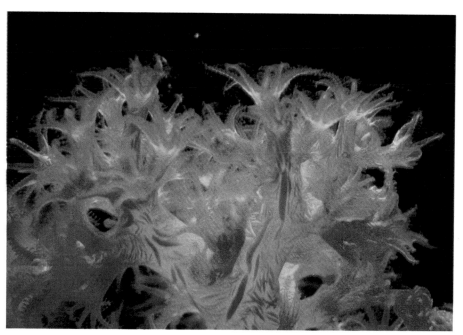

A soft coral from the Red Sea poses the question 'is it a colony or is it a single animal?' Each of the star-like rings of stinging tentacles belongs to a hydra-like animal called a polyp, but all the polyps are interconnected by a canal system and share food materials. The orange speckles are calcium spicules that help support the colony. Although they have no rigid skeletons, some of these corals can reach diameters of over a metre.

Further reading

ALEXANDER, R. McN.: *Animal Mechanics*, Sidgwick & Jackson, 1968 (2nd edition in preparation)
– *The Chordates*, Cambridge University Press, 1975
BELLAIRS, A. D'A and ATTRIDGE, J.: *Reptiles*, Hutchinson University Library, 1975
BLANEY, W.: *How Insects Live*, Elsevier Phaidon, 1976
BURTON, R.: *The Language of Smell*, Routledge & Kegan Paul, 1976
CHAPLIN, R. E.: *Deer*, Blandford, 1977
COLINVAUX, P.: *Why Big Fierce Animals are Rare*, Penguin Books, Harmondsworth, 1980
DALTON, S.: *The Miracle of Flight*, Sampson Low, 1977
GOULD, S. J.: *Ever Since Darwin*, Penguin Books, 1980
HARDY, A. C.: *The Open Sea: The World of Plankton*, Collins, 1956
HILDEBRAND, M.: *Analysis of Vertebrate Structure*, John Wiley & Sons, 1974
HYMAN, L.: *The Invertebrates: Mollusca*, McGraw Hill, 1967
JUDSON, H. F.: *The Search for Solutions*, Hutchinson, 1980
LOCKLEY, R. M.: *Whales, Dolphins and Porpoises*, David and Charles, 1979
MATTHEWS, L. H.: *The Life of Mammals*, Weidenfeld & Nicholson, 1971
MATTHEWS, L. H. and KNIGHT, M.: *The Senses of Animals*, Museum Press, 1963
MILNE, L. J. & M.: *The Senses of Animals and Men*, Andre Deutsch, 1963
NAPIER, J.: *Hands*, George Allen & Unwin, 1980

NORMAN, J. R. (edited by P. H. Greenwood): *A History of Fishes*, 3rd edition, Benn, 1975
PARKS, P.: *The World You Never See: Underwater Life*, Hamlyn, 1976
RICHARDS, O. W. and DAVIES, R. G.: *General Textbook of Entomology*, 10th edition, Chapman & Hall, 1977
SCHMIDT-NIELSEN, K.: *How Animals Work*, Cambridge University Press, 1972
SCHWENK, T.: *Sensitive Chaos*, Rudolf Steiner Press, 1965
SMYTHE, R. H.: *Vision in the Animal World*, Macmillan, 1975
STEELE, R. and HARVEY, A. (editors): *The Encyclopedia of Prehistoric Life*, Mitchell Beazley, 1979
STEVENS, P. S.: *Patterns in Nature*, Penguin Books, 1976
THOMPSON, D'A.: *On Growth and Form*, abridged edition edited by J. T. Bonner, Cambridge University Press, 1961
WELLS, M.: *Lower Animals*, Weidenfeld & Nicholson (World University Library), 1968
WELTY, J. C.: *The Life of Birds*, Constable, 1964
WHITFIELD, P. (editor): *The Animal Family*, Hamlyn, 1976
WHYTE, L. L. (editor): *Aspects of Form*, Lund Humphries, 1968
YONGE, C. M. and THOMPSON, T. E.: *Living Marine Molluscs*, Collins, 1976
YOUNG, J. Z.: *The Life of Mammals*, Oxford University Press, 1957
– *The Life of Vertebrates*, Oxford University Press, 1950

Two series of short, cheap, specialist books are available: the Institute of Biology series, called Studies in Biology (SIB), published by Edward Arnold, and the Oxford/Carolina Biology Readers (O/CBR), published by the Carolina Biological Supply Company and Packard Publishing Limited.

Titles in the Studies in Biology series which deal with topics covered by this book are:

ALEXANDER, R. MC.: *Size and Shape*, SIB 29, 1971
AVERY, R. A.: *Lizards, a Study in Thermoregulation*, SIB 109, 1979
CHAPMAN, R. F.: *A Biology of Locusts*, SIB 71, 1976
CURREY, J. D.: *Animal Skeletons*, SIB 22, 1970
FREE, J. B.: *The Social Organization of Honeybees*, SIB 81, 1977
NEVILLE, A. C.: *Animal Asymmetry*, SIB 67, 1976
PENNYCUICK, C. J.: *Animal Flight*, SIB 33, 1972
RYDER, M. L.: *Hair*, SIB 41, 1973
STODDART, D. M.: *Mammalian Odours and Pheromones*, SIB 73, 1976
WELLS, R. M. G.: *Invertebrate Respiration*, SIB 127, 1980
WICKSTEAD, J. H.: *Marine Zooplankton*, SIB 62, 1976
WILKIE, D. R.: *Muscle*, SIB 11, 1976

In the Oxford/Carolina Biology Readers the following titles are appropriate:

BEER, G. DE: *The Evolution of Flying and Flightless Birds*, OBR 68, 1975
BONE, Q.: *The Origin of the Chordates*, (2nd edition), CBR 18, 1979
BREATHNACH, A. S.: *Melanin Pigmentation of the Skin*, OBR 7, 1971
DENTON, E. J.: *Buoyancy in Marine Animals*, OBR 54, 1974
HARRISON, R. J. and KOOYMAN, G. L.: *Diving in Marine Mammals*, (2nd edition), CBR 6, 1982
MILES, A. E. W.: *Teeth and their Origins*, OBR 21, 1972
NAPIER, J. R.: *Primate Locomotion*, OBR 41, 1976
– *The Human Hand*, CBR 61, 1976
– *Primates and their Adaptations*, (2nd edition), CBR 28, 1977
NEVILLE, A. C.: *The Arthropod Cuticle*, CBR 103, 1978
OSTROM, J. H.: *Dinosaurs*, CBR 98, 1981
PRINGLE, J. W. S.: *Insect Flight*, OBR 52, 1975
PRITCHARD, J. J.: *Bones*, (2nd edition), CBR 47, 1979
WIDDOWSON, E. M.: *Feeding the Newborn Mammal*, CBR 112, 1981
WIGGLESWORTH, V.: *Insect Respiration*, OBR 48, 1972

Acknowledgements

THE INSPIRATION for writing this book has come from many sources, not least from the observation of inspired photographers and from the experience of distinguished zoologists.

Many people read all or part of the manuscript and helped with the illustrations and production. Grateful thanks are due to R. McNeill Alexander, Jill Bailey, Henry Bennet-Clark, George Bernard, Elwyn Blacker, Michael Blacker, John Cooke, Stephen Dalton, Vanessa Hamilton, Michael Packard, Gerald Thompson, Rosemary Vane-Wright and Ronald Ydenberg.

Picture credits

THE PHOTOGRAPHS in this book were supplied by Oxford Scientific Films except those shown /Animals Animals which were supplied by Oxford Scientific Films in association with Animals Animals.

FRONT COVER: George Bernard. BACK COVER: Stephen Dalton. HALF-TITLE: Lynn M. Stone/Animals Animals. TITLE-PAGE: J. C. Stevenson/Animals Animals. ENDPAPER: George Bernard.

INTRODUCTION
8 Stephen Dalton. 9 Stephen Dalton. 10 George Bernard. 11 Stephen Dalton. 12 C. M. Perrins. 13 Stephen Dalton. 14 John Cooke. 15 Peter Parks. 17 Rudie H. Kuiter.

CHAPTER ONE: GEOMETRY OF LIFE
18 Stephen Dalton. 19 TL Peter Parks. BL Stephen Dalton. R Peter Parks. 20 T Peter Parks. B Laurence Gould. 21 T George Bernard. B John Cooke. 22 T Stephen Dalton. B George Bernard. 23 TL George Bernard. TR John Cooke. 24 TR George Bernard. BL Stephen Dalton. 25 TL George Bernard. TR George Bernard. CL John Cooke. CR Peter Parks. B George Bernard. 26 TL George Bernard. TR George Bernard. B John Cooke. 27 T John Cooke. B George Bernard. 28 BL Stephen Dalton. BR L. L. T. Rhodes/Animals Animals. 29 TL Breck P. Kent/Animals Animals. TR Leonard Lee Rue III/Animals Animals. BL George Bernard. BR Mantis Wildlife Films. 30 Stephen Dalton. 31 T Stephen Dalton. B David Thompson. 32 Stephen Dalton. 33 George Bernard. 34 T Stephen Dalton. B Charles Palek. 35 T George Bernard. B George Bernard. 36 Stephen Dalton. 37 TL M. Austerman/Animals Animals. TR Michael Fogden. BL John Cooke. BR John Cooke. 38 TL Laurence Gould. TR Z. Leszczynski/Animals Animals. BL Godfrey Merlen. BR George Bernard. 39 T S. R. Warman. B Sally Foy.

CHAPTER TWO: LIFE SIZE
40 Stephen Dalton. 41 TL David Thompson. BL George Bernard. BC Stephen Dalton. BR Stephen Dalton. 42 TR Stephen Dalton. BL Robert Chapman. BR John Cooke. 43 TL Stephen Dalton. BL Leonard Lee Rue III/Animals Animals. BR Kojo Tanaka/Animals Animals. 44 T Kojo Tanaka/Animals Animals. B Stephen Dalton. 45 Stephen Dalton. 46 TL Tex Fuller/Animals Animals. TR Marty Stouffer/Animals Animals. BL Shep Abbott/Animals Animals. BR Shep Abbott/Animals Animals. 47 T Stephen Dalton. B Margot Conte/Animals Animals. 48 George Bernard. 49 T Raymond A. Mendez/Animals Animals. B John Cooke. 50 T John Paling. BL Stephen Dalton. BR John Cooke. 51 P. & W. Ward. 52 Peter David/Seaphot. 53 John Cooke. 54 T John Cooke. B P. & W. Ward. 55 T John Paling. B S. D. Halperin/Animals Animals. 56 T C. M. Perrins. B George Bernard. 57 T John Cooke. B George Bernard. 58 T Leonard Lee Rue III/Animals Animals. B M. Austerman/Animals Animals. 59 Leonard Lee Rue III/Animals Animals. 60 T John Cooke. BL Z. Leszczynski/Animals Animals. BR David C.

Fritts/Animals Animals. 61 T Breck P. Kent/Animals Animals. B Carl Roessler/Animals Animals. 62 T Peter Parks. B Manfred Kage. 63 Manfred Kage. 64 Peter Parks. 65 Peter Parks. 66 T Leonard Lee Rue III/Animals Animals. B P. C. Lack. 67 Charles Palek/Animals Animals. 68 T M. J. Coe. B Z. Leszczynski/Animals Animals. 69 Animals Animals.

CHAPTER THREE: ON THE SURFACE
70 Stephen Dalton. TR Stephen Dalton. BL Stephen Dalton. BR Sean Morris. 71 Stephen Dalton. 72 George Bernard. 73 T Stephen Dalton. B David Scharf. 74 TL Peter Parks. TR George Bernard. BR John Cooke. 75 T John Cooke. B Mantis Wildlife Films. 76 T Jerry Cooke. B Richard Kolar/Animals Animals. 77 Stephen Dalton. 78 Z. Leszczynski/Animals Animals. 79 TL George Bernard. TR Richard Kolar/Animals Animals. BL Sally Foy. BR M. Austerman/Animals Animals. 80 T M. P. L. Fogden. B Z. Leszczynski/Animals Animals. 81 T George Bernard. B Z. Leszczynski/Animals Animals. 82 BL David Fritts/Animals Animals. BR Stephen Dalton. 83 BL Stouffer Productions/Animals Animals. BR Stephen Dalton. 84 Leonard Lee Rue III/Animals Animals. 85 P. K. Sharpe. 86 George Bernard. 87 Stephen Dalton. 88/89 Rudie H. Kuiter. 90 Peter Parks. 91 Rudie H. Kuiter. 92 T M. P. L. Fogden. B George Bernard. 93 Stephen Dalton. 94 Stephen Dalton. 95 Stephen Dalton. 96 Stephen Dalton. 97 Roger B. Minkoff/Animals Animals. 98 Manfred Kage. 99 BL George Bernard. BR John Cooke. 100 T George Bernard. B Stephen Dalton. 101 Stephen Dalton. 102 D. J. Stradling. 103 John Cooke. 104/5 Stephen Dalton. 106 George Bernard. 107 T George Bernard. BL David Thompson. BR D. J. Stradling. 108 Stephen Dalton. 109 Stephen Dalton. 110 Sally Foy. 111 TL Al Szabo/Animals Animals. TR Stephen Dalton. B Stephen Dalton. 112 T Robert Chapman. B Peter Parks. 113 Peter Parks. 114 T Rudie H. Kuiter. B Alison Kuiter. 115 Stephen Dalton. 116 Peter Parks. 117 T George Bernard. B Peter Parks.

CHAPTER FOUR: FACING UP TO NATURE
118 George Bernard. 119 John Cooke. 120/121 Z. Leszczynski/Animals Animals. 122 Stephen Dalton. 123 David Scharf. 124 David Thompson. 125 T George Bernard. BL Stephen Dalton. BR John Cooke. 126 Stephen Dalton. 127 T Joe McDonald/Animals Animals. B P. & W. Ward. 128 TL John Cooke. TR M. P. L. Fogden. CL Stephen Dalton. CR P. & W. Ward. BL Stephen Dalton. BR John Cooke. 129 David Scharf. 130 TL Stephen Dalton. TR David Thompson. B Stephen Dalton. 131 Stephen Dalton. 132 T Bob Fredrick. B Stephen Dalton. 133 David Scharf. 134 Stephen Dalton. 135 George Bernard. 136 T John Paling. B M. Austerman/Animals Animals. 137 TL L. T. Rhodes/Animals Animals. B Stouffer Productions/Animals Animals. 138 Stouffer Enterprises/Animals Animals. 139 T Z. Leszczynski/Animals Animals. BL David Thompson. BR Stephen Dalton. 140 T Breck P. Kent/Animals Animals. B M. Austerman/Ani-

229

mals Animals. 141 Stephen Dalton. 142 ᴛJohn Cooke. ʙBreck P. Kent/Animals Animals. 143 Stephen Dalton. 144 Stephen Dalton. 145 ᴛJim Doran/Animals Animals. ʙRudie H. Kuiter. 146 Stephen Dalton. 147 Brian Milne/Animals Animals. 148 Robin Buxton. 149 Stephen Dalton. 150 Graham J. Wren. 151 M. Austerman/Animals Animals. 152/153 Stephen Dalton. 154 ᴛʟStephen Dalton. ᴛʀJohn Chellman/Animals Animals. ʙPerry D. Slocum/Animals Animals. 155 ᴛStephen Dalton. ʙRobert Fields/Animals Animals. 156 Stephen Dalton. 157 ᴛʟLynn M. Stone/Animals Animals. ᴛʀM. Austerman/Animals Animals. ʙM. Austerman/Animals Animals. 158 George Bernard. 159 ᴛJohn Cooke. ʙGeorge Bernard. 160 ᴛJohn Cooke. ʙIan G. Moar. 161 ᴛJohn Cooke. ʙStephen Dalton. 162 David Scharf. 163 ᴛStephen Dalton. ʙJohn Cooke. 164 George Bernard. 165 ᴛBreck P. Kent/Animals Animals. ʙM. P. L. Fogden. 166 ᴛM. J. Coe. ʙJohn Cooke. 167 Sally Foy. 168/169 M. J. Coe. 170 Leonard Lee Rue III/Animals Animals. 171 Stephen Dalton.

CHAPTER FIVE: FINISHING TOUCHES
172 John Paling. 173 Esao Hashimoto/Animals Animals, 174 ᴛʟSally Foy. ᴛʀM. Austerman/Animals Animals. ʙStephen Dalton. 175 M. A. Chappell/Animals Animals. 176 ᴛM. J. Coe. ʙʟDavid Thompson. ʙʀSally Foy. 177 John Chellman/Animals Animals. 178 ᴛ Raymond A. Mendez/Animals Animals. ʙA. G. (Bert) Wells. 179 David Thompson. 180 Stephen Dalton. 181 Stephen Dalton. 182 Hans & Judy Beste/Animals Animals. 183 Stephen Dalton. 184 Stephen Dalton. 185 Stephen Dalton. 186 Peter Parks. 187 ᴛGerald Thompson. ʙʟPeter Parks. ʙʀRudie H. Kuiter. 188 ᴛAlison Kuiter. ʙPeter Parks. 189 Peter Parks. 190 George Bernard. 191 ᴛAndrew Lister. ʙJohn Paling. 192 Z. Leszczynski/Animals Animals. 193 Laurence Gould. 194 Z. Leszczynski/Animals Animals. 195 Z. Leszczynski/Animals Animals. 196 Stephen Dalton. 197 Stephen Dalton. 198 ᴛLynn M. Stone/Animals Animals. ʙP. & W. Ward. 199 ᴛStephen Dalton. ʙGraham J. Wren. 200/201 Stephen Dalton. 202 Stephen Dalton. 203 Stephen Dalton. 204 Stephen Dalton. 205 Jack Wilburn/Animals Animals. 206 ᴛLen Rue Jnr./Animals Animals. ʙDavid Thompson. 208 M. J. Coe. 209 George Bernard. 210 Leonard Lee Rue III/Animals Animals. 211 Leonard Lee Rue III/Animals Animals. 212 George Bernard. 213 Z. Leszczynski/Animals Animals. 214 Stephen Dalton. ʙJohn Paling. 215 ᴛLeonard Lee Rue III/Animals Animals. ʙA. G. (Bert) Wells. 216/217 Stephen Dalton. 218 Stephen Dalton. 219 ᴛGeorge Bernard. ʙCharles Palek/Animals Animals. 220 ᴛGeorge Bernard. ʙP. & W. Ward. 221 ᴛJohn Cooke. ʙM. Austerman/Animals Animals. 222 Stephen Dalton. 223 George Bernard. 224 Peter Parks. 225 Laurence Gould.

Index